SPRINGER PROTOCOLS HANDBOOKS

For further volumes:
http://www.springer.com/series/8623

Springer Protocols Handbooks collects a diverse range of step-by-step laboratory methods and protocols from across the life and biomedical sciences. Each protocol is provided in the Springer Protocol format: readily-reproducible in a step-by-step fashion. Each protocol opens with an introductory overview, a list of the materials and reagents needed to complete the experiment, and is followed by a detailed procedure supported by a helpful notes section offering tips and tricks of the trade as well as troubleshooting advice. With a focus on large comprehensive protocol collections and an international authorship, Springer Protocols Handbooks are a valuable addition to the laboratory.

Immunity in Insects

Edited by

Federica Sandrelli

Department of Biology, University of Padova, Padova, Italy

Gianluca Tettamanti

Department of Biotechnology and Life Sciences,
University of Insubria, Varese, Italy

Humana Press

Editors
Federica Sandrelli
Department of Biology
University of Padova
Padova, Italy

Gianluca Tettamanti
Department of Biotechnology and Life Sciences
University of Insubria
Varese, Italy

ISSN 1949-2448 ISSN 1949-2456 (electronic)
Springer Protocols Handbooks
ISBN 978-1-0716-0258-4 ISBN 978-1-0716-0259-1 (eBook)
https://doi.org/10.1007/978-1-0716-0259-1

This Humana imprint is published by the registered company Springer Science+Business Media, LLC, part of Springer Nature.
The registered company address is: 233 Spring Street, New York, NY 10013, U.S.A.

Preface

Immunity in Insects as a field of research has been in constant expansion over the past 30 years. Indeed, a Pubmed search using "insect immunity" as a keyword shows that the total number of published papers on this topic almost tripling every decade. The identification of the first antimicrobial peptide Cecropin from the lepidopteran *Hyalophora cecropia* by Boman in 1981 is considered the seminal work, which paved the way for this research field. The later clarification of immune response regulation and activation of innate immunity in the model organism *Drosophila melanogaster* represents a fundamental milestone that led Jules Hoffmann to be awarded the Nobel prize in Physiology or Medicine in 2011. Since then the immune response in insects has become a multidisciplinary field, and currently the study of host–pathogen relationships in these animals appears fundamental in elucidating evolutionary and ecological traits, in controlling both beneficial and agriculture-pest insects, as well as in the study of insect vectors of tropical diseases. Although insects lack an adaptive immune response, their innate immune response displays remarkable homologies with that of vertebrates. Additionally, insects often have short life cycles and produce large numbers of individuals each generation. For these reasons, in recent years several dipteran and lepidopteran species have been recognized as cost-effective model organisms to study human microbial infections, implementing the "3R principle" of animal use. Insects also represent a significant source of antimicrobial peptides (AMPs), molecules nowadays analyzed as potential alternatives to conventional antibiotics to treat infections, and also for their immunomodulatory function and anticancer activity.

This book aims at providing a methodological guide for scientists interested in different aspects of insect immunity. In the first part, Chapters 1 and 2 report up-to-date genomic and transcriptomic approaches to study the immune response and identify new immune genes and proteins in both model and non-model insect species, with a step-by-step experiment description. Chapter 3 provides a detailed methodology to prepare insect samples for proteomic analysis, a powerful technique in understanding protein functions and networks of the insect immune system. The second part reports fundamental techniques to induce, characterize, and monitor cellular (Chapters 4–6) and humoral (Chapters 7 and 8) immune responses. Chapter 4 describes assays for the absolute quantification of hemocytes in *Drosophila* and other insects, while Chapter 5 provides protocols to study *in vivo* and *ex vivo* phagocytosis as well as the opsonization process in *Drosophila*. Chapter 6 reports an overview of methods for the analysis of the cellular response in Lepidoptera to assess hemocyte behavior in larvae affected by pathogens and parasitoids. Both Chapters 7 and 8 are focused on Lepidoptera but, while the first presents a broad array of approaches for detecting the distribution, expression, and activity of Phenoloxidase, the second goes beyond melanization and Prophenoloxidase system and considers other mechanisms of the humoral response such as lysozyme and AMPs. Chapter 9 then focuses on insect AMPs as possible alternatives to conventional antibiotics, providing methodologies to analyze their *in vitro* antimicrobial activity and potential toxicity against human cells. Chapter 10 describes molecular dynamic simulation approaches to elucidate the interaction between AMPs and bacterial membranes. In recent years, the microbiota has gained attention as a fundamental factor in modulating several aspects of the host physiology, including development and immunity. Focusing on the mosquito *Aedes aegypti*, the main vector of several arboviruses, Chapter 11 reports protocols to manipulate the composition of mosquito microbiota to investigate its effect on host physiology and vector competence. Chapter 12 focuses on viral infections in insects, providing a bioinformatic pipeline to identify RNA viruses and the antiviral immune response using RNA sequence data. We chose to dedicate the final part of this book to the use of insects as model organisms to study infections. Chapter 13 describes methodologies to investigate infections by entomopathogenic nematodes that are effective biocontrol agents against insect pests and monitor the immune response in *Drosophila*; Chapter 14 presents protocols to analyze the immune response following oral infections in the model insect *Bombyx mori* and Chapter 15 provides protocols to monitor the effect of septic infections with human pathogens using *B. mori* as a model.

Padova, Italy — *Federica Sandrelli*
Varese, Italy — *Gianluca Tettamanti*

Contents

Contributors

ANDREA BECCHIMANZI • *Department of Agriculture, Università degli Studi di Napoli "Federico II", Portici, NA, Italy*

ANIRBAN BHUNIA • *Department of Biophysics, Bose Institute, Kolkata, India*

DANIEL BRADY • *Department of Biology, University of Padova, Padova, Italy*

MAURIZIO FRANCESCO BRIVIO • *Lab. of Comparative Immunology and Parasitology, Department of Theoretical and Applied Sciences, University of Insubria, Varese, Italy*

KATJA BRÜCKNER • *Eli and Edythe Broad Center of Regeneration Medicine and Stem Cell Research, University of California San Francisco, San Francisco, CA, USA; Department of Cell and Tissue Biology, University of California San Francisco, San Francisco, CA, USA; Cardiovascular Research Institute, University of California San Francisco, San Francisco, CA, USA*

SILVIA CACCIA • *Department of Agriculture, Università degli Studi di Napoli "Federico II", Portici, NA, Italy*

SILVIA CAPPELLOZZA • *Council for Agricultural Research and Economics, Research Centre for Agriculture and Environment, Sericulture Laboratory, Padova, Italy*

FLORIANA CAPPIELLO • *Laboratory affiliated to Pasteur Italia-Fondazione Cenci Bolognetti, Department of Biochemical Sciences "A. Rossi Fanelli", Sapienza University of Rome, Rome, Italy*

BRUNO CASCIARO • *Laboratory affiliated to Pasteur Italia-Fondazione Cenci Bolognetti, Department of Biochemical Sciences "A. Rossi Fanelli", Sapienza University of Rome, Rome, Italy; Center for Life Nano Science, Istituto Italiano di Tecnologia, Rome, Italy*

SEAN CORCORAN • *Department of Cell and Tissue Biology, University of California San Francisco, San Francisco, CA, USA*

ILARIA DI LELIO • *Department of Agriculture, Università degli Studi di Napoli "Federico II", Portici, NA, Italy*

IOANNIS ELEFTHERIANOS • *Infection and Innate Immunity Lab, Department of Biological Sciences, The George Washington University, Washington, DC, USA*

DOMINIQUE FERRANDON • *Sino-French Hoffmann Institute, Guangzhou Medical University, Guangzhou, P. R. China; Université de Strasbourg, CNRS, Strasbourg, France*

MATHILDE GENDRIN • *Institut Pasteur de la Guyane, Cayenne, French Guiana, France; Institut Pasteur, Paris, France*

SHULIN HE • *Institute of Biology, Freie Universität Berlin, Berlin, Germany; Department for Materials and Environment, BAM Federal Institute for Materials Research and Testing, Berlin, Germany*

CHRISTA HERYANTO • *Infection and Innate Immunity Lab, Department of Biological Sciences, The George Washington University, Washington, DC, USA*

WUREN HUANG • *Key Laboratory of Insect Developmental and Evolutionary Biology, CAS Center for Excellence in Molecular Plant Sciences, Shanghai Institute of Plant Physiology and Ecology, Chinese Academy of Sciences, Shanghai, China*

PAUL R. JOHNSTON • *Institute of Biology, Freie Universität Berlin, Berlin, Germany; Berlin Center for Genomics in Biodiversity Research, Berlin, Germany; Leibniz-Institute of Freshwater Ecology and Inland Fisheries (IGB), Berlin, Germany*

JAANA JURVANSUU • *Ecology and Genetics Research Unit, University of Oulu, Oulu, Finland*

RAJIV K. KAR • *Department of Biophysics, Bose Institute, Kolkata, India*

ERIC KENNEY • *Infection and Innate Immunity Lab, Department of Biological Sciences, The George Washington University, Washington, DC, USA*

BRIAN P. LAZZARO • *Cornell Institute of Host-Microbe Interactions and Disease, Departments of Entomology and Ecology & Evolutionary Biology, Cornell University, Ithaca, NY, USA*

SAMUEL LIÉGEOIS • *Sino-French Hoffmann Institute, Guangzhou Medical University, Guangzhou, P. R. China; Université de Strasbourg, CNRS, M3I UPR 9022, Strasbourg, France*

MARIA ROSA LOFFREDO • *Laboratory affiliated to Pasteur Italia-Fondazione Cenci Bolognetti, Department of Biochemical Sciences "A. Rossi Fanelli", Sapienza University of Rome, Rome, Italy*

YUFA LUO • *College of Life Sciences, Shangrao Normal University, Shangrao, China*

MARIA LUISA MANGONI • *Laboratory affiliated to Pasteur Italia-Fondazione Cenci Bolognetti, Department of Biochemical Sciences "A. Rossi Fanelli", Sapienza University of Rome, Rome, Italy*

MARISTELLA MASTORE • *Lab. of Comparative Immunology and Parasitology, Department of Theoretical and Applied Sciences, University of Insubria, Varese, Italy*

YASUHIKO MATSUMOTO • *Department of Microbiology, Meiji Pharmaceutical University, Tokyo, Japan; Teikyo University Institute of Medical Mycology, Tokyo, Japan*

DINO P. MCMAHON • *Institute of Biology, Freie Universität Berlin, Berlin, Germany; Department for Materials and Environment, BAM Federal Institute for Materials Research and Testing, Berlin, Germany*

SHRUTI MUKHERJEE • *Department of Biophysics, Bose Institute, Kolkata, India*

YUYANG NI • *College of Life Sciences, Shangrao Normal University, Shangrao, China*

FRANCESCO PENNACCHIO • *Department of Agriculture, Università degli Studi di Napoli "Federico II", Portici, NA, Italy*

ALEXANDRO RODRÍGUEZ-ROJAS • *Institute of Biology, Freie Universität Berlin, Berlin, Germany*

JENS ROLFF • *Institute of Biology, Freie Universität Berlin, Berlin, Germany*

OTTAVIA ROMOLI • *Department of Biology, University of Padova, Padova, Italy; Institut Pasteur de la Guyane, Cayenne, French Guiana, France*

TIMOTHY B. SACKTON • *Faculty of Arts and Sciences, Harvard University, Cambridge, MA, USA*

FEDERICA SANDRELLI • *Department of Biology, University of Padova, Padova, Italy*

ALESSIO SAVIANE • *Council for Agricultural Research and Economics, Research Centre for Agriculture and Environment, Sericulture Laboratory, Padova, Italy*

KAZUHISA SEKIMIZU • *Teikyo University Institute of Medical Mycology, Tokyo, Japan; Genome Pharmaceuticals Institute Co. Ltd., Tokyo, Japan*

GIANLUCA TETTAMANTI • *Department of Biotechnology and Life Sciences, University of Insubria, Varese, Italy*

LUMI VILJAKAINEN • *Ecology and Genetics Research Unit, University of Oulu, Oulu, Finland*

Jing Wang • *College of Life Sciences, Shangrao Normal University, Shangrao, China*
Wenhui Wang • *Sino-French Hoffmann Institute, Guangzhou Medical University, Guangzhou, P. R. China; Université de Strasbourg, CNRS, M3I UPR 9022, Strasbourg, France*
Robert M. Waterhouse • *Department of Ecology and Evolution and Swiss Institute of Bioinformatics, University of Lausanne, Lausanne, Switzerland*
Kai Wu • *College of Life Sciences, Shangrao Normal University, Shangrao, China*
Bing Yang • *Key Laboratory of Insect Developmental and Evolutionary Biology, CAS Center for Excellence in Molecular Plant Sciences, Shanghai Institute of Plant Physiology and Ecology, Chinese Academy of Sciences, Shanghai, China*

Part I

Searching for Genes and Proteins Involved in the Immune Response

Chapter 1

Characterization of Insect Immune Systems from Genomic Data

Robert M. Waterhouse, Brian P. Lazzaro, and Timothy B. Sackton

Abstract

Insects face a multitude of threats from the pathogens and parasites they encounter over their life cycles, and they use robust immune systems to defend themselves. This chapter provides a tutorial for the identification and annotation of genes that comprise the immune system from newly sequenced insect genomes. Insect immune responses are orchestrated by the products of a suite of genes responsible for pathogen recognition, signal transduction, and pathogen killing. Many of the genes and proteins underlying these processes can be identified based on sequence homology with related species that have been immunologically characterized. Additional components of the immune response can be identified by transcriptomic analyses to detect genes whose expression changes in response to infection stimulus. Application of our step-by-step protocols for these complementary approaches enables the characterization of insect immune systems from genomic data.

Key words Immunity, Infection, Genome annotation, Gene families, Comparative genomics, Transcriptomics

1 Introduction

A major element of genome sequencing projects is the identification and annotation of the genes expected to underlie key physiological processes. The initial identification of these genes from genomic data enables subsequent functional experimentation and comparative genomic analyses to understand the evolutionary forces that drive establishment, maintenance, and diversification of these processes. In this chapter, we describe (1) a general framework for using sequence homology searches and (2) a detailed infection protocol for transcriptomic analyses to identify and annotate candidate immune system genes in newly sequenced insect genomes.

The identification of genes in newly sequenced genomes is typically initiated with computational searches for homologs of

Federica Sandrelli and Gianluca Tettamanti (eds.), *Immunity in Insects*, Springer Protocols Handbooks,
https://doi.org/10.1007/978-1-0716-0259-1_1,

genes that have been characterized in other species. This approach works well for genes that make up an evolutionarily conserved, canonical immune repertoire, such as those established over two decades of functional genetic research on the model insect *Drosophila melanogaster* [1–6] and more recent work in non-model insects [7–16]. The identification of novel genes or those with no prior ascribed functional role in immunity, however, requires experimental data to be coupled with the computational analyses. Identifying these infection-responsive genes is facilitated by the fact that the expression of many immune genes is induced by infectious challenge. This means that transcriptomic analysis of changes in gene expression after infection can be used to support inferences from homology searches and to suggest additional, sometimes novel, components of the immune system.

Homology searches are excellent for identifying conserved genes and protein domains that comprise various components of the innate immune system. This includes most immune gene families and signaling pathway members. The presence of core recognition, signaling and modulation, and effector components of the immune system indicates functional conservation across taxa, while notable absences such as the apparent degradation of the IMD pathway in pea aphids [10] can suggest possible rewiring of the system. Computational searches will identify candidate immune-related genes from the full set of genes predicted by whole genome annotation pipelines. Manual curation may be required to validate some candidates or to confirm cases of apparent losses of otherwise widely conserved genes. Homology searches also help to detect and quantify expansions and contractions of multi-gene families that vary in copy number across insects, such as genes encoding peptidoglycan recognition proteins (PGRPs) and members of the phenoloxidase cascade (PPOs). Unlike for the generally single-copy signaling pathway genes, defining clear orthologous relationships can be difficult for such multi-gene families, depending on the age of the gene duplications and the phylogenetic distance between the species being compared. Nevertheless, the variable numbers of such immunity genes can sometimes be interpreted as indicative of the natural selective and epidemiological pressures on the insect being studied [7, 17, 18].

Homology searches are invaluable for identifying most canonical immune genes. However, genes that have newly acquired immune functions, or evolutionarily novel genes with roles in immunity, will not be identified through homology searches using known immune gene sequences. Thus homology searches can be complemented with transcriptomic analyses to identify sets of genes whose expression levels are responsive to infection, but that are not normally considered part of the canonical immune system. In such analyses, the insect in question is challenged with a relevant infection stimulus and RNA is extracted either from the whole insect or from immunologically relevant tissues. The gene expression profiles of challenged insects can then be compared to the expression profiles of naïve insects, enabling identification of genes whose expression is induced or repressed by infection (e.g., [19, 20]). Transcriptomic analysis is especially powerful for identifying effector genes such as those encoding antimicrobial peptides (AMPs). These may be unique to specific groups of insects and the genes are often so short that they fail to be detected by computational gene-finding algorithms. However, they are often massively transcriptionally induced upon infection. Thus, transcriptomic analysis can be a powerful approach to identify effectors that would be missed by other methods (reviewed in [21]). While AMPs and other effectors have direct roles in immunity, many other differentially expressed genes may play indirect roles, and as such they do not form part of the "immune system" by any canonical definition. For example, infection often causes activation of generic stress response genes [22, 23] and a transcriptional signature of repression of basal metabolism [24, 25]. In some cases, these transcriptional responses may promote host survival, but in other cases they may even represent deleterious consequences of infection. Therefore caution must be taken and it should not be assumed that a gene is part of the immune system solely because its expression level changes after challenge.

Homology searches and transcriptomic analyses are complementary approaches to characterize genes that play a role in the insect immune system from newly sequenced genomes (henceforth referred to as the "target" or "focal" species). Sequence homology searching is powerful and allows for the identification of genes with conserved immune-related protein domains, including genes whose expression patterns do not change substantially in response to infections. Transcriptomic analyses have the advantage that they can identify novel infection-responsive genes that have not been previously characterized in other species. In this chapter, we detail a practical workflow for applying these two approaches in parallel to characterize the immune system of an insect with a newly sequenced genome.

2 Methods

2.1 Identification of Canonical Innate Immunity Genes

Characterizing the canonical innate immune gene repertoire in newly sequenced genomes follows four main steps, presented in Fig. 1. The first is to compile a comprehensive list of immune-related genes and their protein sequences from species that have already been characterized (henceforth referred to as the "reference" species). These sequences are then used to search the

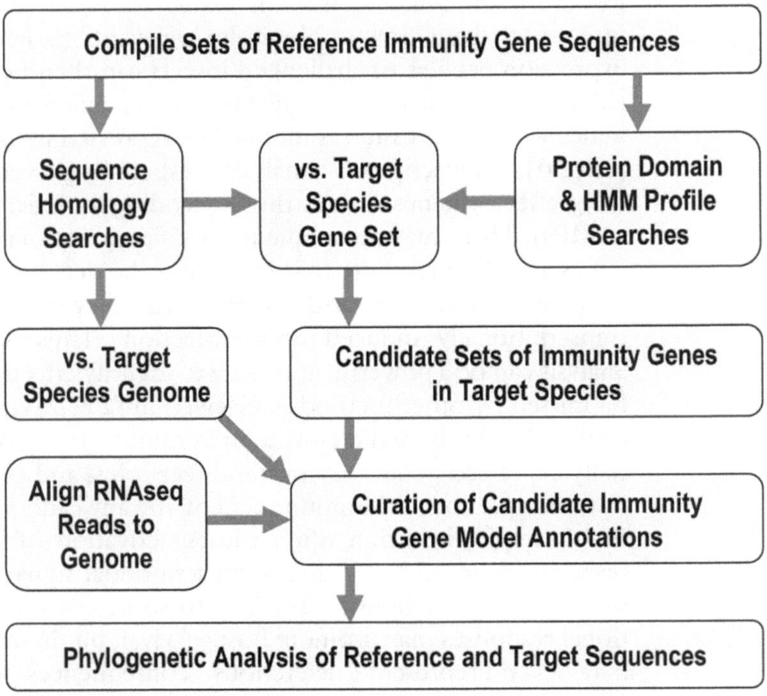

Fig. 1 Workflow of steps required for canonical immune gene identification. Protein sequences of immune-related genes from selected reference species are first collected based on the current knowledge of insect innate immunity. These are then used as reference query sequences and sequence hidden Markov model (HMM) profiles for homology searches of the gene set (protein sequences) of the target species to be investigated. Complementary protein-domain searches are used to identify genes that contain domains in common with the reference immunity genes. Results from the sequence and domain searches are then used to prioritize the inspection of the candidate immunity genes and curate their predicted gene models to ensure they are as complete and accurate as possible. This will benefit from the results from homology searches of the reference query sequences against the genome assembly as well-aligned RNA sequencing (RNA-seq) reads from the target species. Combined phylogenetic analysis of homologous reference and target candidate sequences to build gene trees then allows for the confirmation or rejection of the candidate immune-related genes and the characterization of their orthologous or paralogous relationships

genomes and gene sets (the complete set of predicted genes for a given genome) for putative homologs and characteristic protein domains. The candidate gene models can then be inspected and manually curated to ensure that they are correct and complete. Finally, phylogenetic analyses to trace the evolutionary histories of each gene family allow for the delineation of orthologs and paralogs and the confident characterization of a new set of canonical immune genes.

2.1.1 Compiling Sets of Reference Sequences

1. The comparative approach to identifying immune-related genes in newly sequenced genomes relies on comparisons with previously characterized sets of immunity genes in other species. While newer investigations of immune systems across diverse

insect taxa have begun to reveal novelties in different species, a great deal of the collective knowledge of the canonical insect innate immune gene repertoire nevertheless still derives from studies conducted on *D. melanogaster* (*see* **Note 1**). To start compiling sets of reference immune gene sequences, you will first need to (1) define the scope of your study by deciding which immune-related pathways and gene families to include and (2) select appropriate species from which to source the reference immune protein sequences.

2. Defining the scope of the immune gene repertoire to be examined requires an overview of the current understanding of the canonical insect innate immune system. The principal components of an immune response must include proteins responsible for recognition of pathogens, signal transduction once a pathogen has been recognized, and effector proteins and biomolecules that eliminate the pathogen (Table 1). A core set

Table 1
The principal components of the canonical insect innate immune gene repertoire

Gene family or signaling pathway	Brief description
IMD pathway	The immune deficiency pathway is characterized by peptidoglycan recognition protein receptors, intracellular signal transducers and modulators, and the NF-κB transcription factor relish
Toll pathway	The intracellular components of Toll pathway signaling are homologous to the Toll-like receptor innate immune pathway in mammals, culminating in activation of the NF-κB transcription factors dorsal and DIF in *Drosophila*
JAK/STAT pathway	The JAnus kinase protein (JAK) and the signal transducer and activator of transcription (STAT) are two core components of the JAK/STAT pathway, which is involved in cellular responses to stress or injury
RNAi pathway	RNA interference protects against viral infections employing dicer and Argonaute proteins as well as helicases to identify and destroy exogenous double-stranded RNAs
Antimicrobial peptides	AMPs are the classical effector molecules of innate immunity; they include defensins, cecropins, and attacins that are involved in bacterial killing by disrupting their membranes
Caspases	Cysteine-aspartic proteases are involved in immune signaling cascades and apoptosis
CLIP-domain serine proteases	Several CLIP proteases have roles as activators or modulators of immune signaling cascades
C-type lectins	CTLs are carbohydrate-binding proteins with roles in pathogen opsonization, encapsulation, and melanization, as well as immune signaling cascades

(continued)

Table 1
(continued)

Gene family or signaling pathway	Brief description
Fibrinogen-related proteins	FREPs (also known as FBNs) are a family of pattern recognition receptors with homology to the C terminus of the fibrinogen β- and γ-chains
Galectins	GALEs bind specifically to β-galactoside sugars and can function as pattern recognition receptors in innate immunity
Gram-negative binding proteins	GNBPs (or β-1,3-glucan-binding proteins, BGBPs) are a family of carbohydrate-binding pattern recognition receptors
Inhibitors of apoptosis	IAPs are important in antiviral responses and are involved in regulating immune signaling and suppressing apoptotic cell death
Lysozymes	LYSs are key effector enzymes that hydrolyze peptidoglycans present in the cell walls of many bacteria, causing cell lysis
MD-2-like proteins	MLs, also known as Niemann-pick type C-2 proteins, possess myeloid-differentiation-2-related lipid-recognition domains involved in recognizing lipopolysaccharide
Nimrods	NIMs have been shown to bind bacteria leading to their phagocytosis by hemocytes
Peptidoglycan recognition proteins	PGRPs are pattern recognition receptors capable of recognizing the peptidoglycan from bacterial cell walls
Prophenoloxidases	PPOs are key enzymes in the melanization cascade that helps to kill invading pathogens and is important for wound healing
Peroxidases	PRDXs are enzymes involved in the metabolism of reactive oxygen species (ROS) that are toxic to pathogens
Scavenger receptors	SCRs are made up of different classes that function as pattern recognition receptors for a broad range of ligands including from pathogens
Superoxide dismutases	SODs are antioxidant enzymes involved in the metabolism of toxic superoxide into oxygen or hydrogen peroxide
Spaetzle-like proteins	The cleavage of Spaetzle results in binding of the product to the toll receptor and subsequent activation of the toll pathway; SPZs contain a cystine knot domain
Serine protease inhibitors	Protease inhibition by serpins, or SRPNs, modulates many signaling cascades; they act as suicide substrates to inhibit their target proteases
Thioester-containing proteins	TEPs are related to vertebrate complement factors and α2-macroglobulin protease inhibitors; their activation through proteolytic cleavage leads to phagocytosis or killing of pathogens

of key genes and pathways has been characterized through experimental research in different insect systems and shown to be widely conserved across divergent insect species (*see* **Note 2**). These can serve as the initial basis for homology searches, although novel genes should also be expected to emerge from each new study system. A streamlined scope would normally first focus on (1) canonical families of pathogen recognition receptors such as peptidoglycan recognition proteins (PGRPs) and Gram-negative bacteria-binding proteins (GNBPs, also known as beta-1,3-glucan-binding proteins); (2) the core members of the three main immune signaling cascades, the Toll, IMD, and JAK/STAT pathways; and (3) effectors such as antimicrobial peptides (AMPs) and lysozymes (LYSs) whose expression is generally upregulated upon stimulation of these pathways. Additional core processes include immune responses such as RNA interference (RNAi), phagocytosis, apoptosis and autophagy, the defensive production of reactive oxygen species (ROS), and melanization reactions [26, 27]. Broadening the scope of the study further would normally include (1) additional gene families with members implicated in pathogen recognition and/or immune response activation such as C-type lectins (CTLs), thioester-containing proteins (TEPs), or scavenger receptors (SCRs) and (2) genes responsible for the positive or negative regulation of core members of the main signaling pathways and cascade modulation. Ultimately, the scope of the study will be determined by size of the research team working on the project and the questions of particular biological interest for the target species.

3. The selection of appropriate reference species should be guided by published comparative characterizations of other insect genomes such as those listed in Table 2. Selecting several reference species will allow for better consistency checks, e.g., do searches using one reference species produce similar results as using another reference species? Comparisons between insects from the same order are the most useful, as the lower sequence divergence between more closely related species improves the success of sequence homology searches. Additionally, gene family composition will generally be more similar between closely related species, with fewer gene gains or losses since their last common ancestor. Data from the reference species should be public, versioned, and recognized by their respective communities as the official assemblies and gene sets, to facilitate both repeatability of the analysis and ease of data acquisition. Data retrieval and querying will be further facilitated if the selected reference species are already hosted by an online genome browser resource such as the Bioinformatics Platform for Agroecosystem Arthropods [28],

Ensembl Metazoa [29], FlyBase [30], Hymenoptera Genome Database [31], i5k at the National Agricultural Library [32], the National Center for Biotechnology Information [33], or VectorBase [34].

4. Having defined the scope and selected the reference species, you can now proceed with compiling your sets of reference immune-related protein sequences. Published studies such as those presented in Table 2 usually include lists of gene and/or protein identifiers of the immune genes that were identified. Use these to extract the corresponding sequences from the complete gene sets for each species. As these studies are effectively snapshots of the available data at the time of publication,

Table 2
Examples of comparative studies of the canonical insect innate immune repertoire

Focal species	Comparison species	Breadth of study	Reference
Six *Glossina*	*Musca domestica* *Drosophila melanogaster*	Rec, Sig, Mod, Eff	Attardo et al. [35]
Manduca sexta	*Bombyx mori*	Serine protease inhibitors (SRPNs)	Li et al. [36]
Aedes aegypti	*Aedes albopictus* *Anopheles gambiae* *Culex quinquefasciatus*	C-type lectins (CTLs)	Adelman and Myles [37]
Six *Glossina*	Several other dipterans Outgroup blood-feeding hemipterans	Thioester-containing proteins (TEPs)	Matetovici and Van Den Abbeele [38]
Musca domestica	*Glossina morsitans* Five mosquitoes Seven *Drosophila*	Rec, Sig, Mod, Eff	Sackton et al. [7]
Pteromalus puparum	*Aedes aegypti* *Anopheles gambiae* *Apis mellifera* *Bombyx mori* *Drosophila melanogaster* *Manduca sexta*	Serine protease inhibitors (SRPNs)	Yang et al. [39]
Bombus impatiens *Bombus terrestris*	*Apis florea* *Apis mellifera* *Megachile rotundata* *Nasonia vitripennis* *Tribolium castaneum* *Drosophila melanogaster* *Anopheles gambiae*	Rec, Sig, Mod, Eff	Barribeau et al. [8]
Anopheles gambiae	Twenty other mosquitoes *Drosophila melanogaster*	Rec, Sig, Mod, Eff	Neafsey et al. [40]

(continued)

Table 2 (continued)

Focal species	Comparison species	Breadth of study	Reference
Zootermopsis nevadensis	Diptera Lepidoptera Coleoptera Hymenoptera	Rec, Sig, Mod, Eff	Terrapon et al. [41]
Nasonia vitripennis	*Drosophila melanogaster* *Anopheles gambiae* *Apis mellifera* *Acyrthosiphon pisum*	Rec, Sig, Mod, Eff	Brucker et al. [42]
Aedes aegypti	*Anopheles gambiae* *Culex quinquefasciatus* Twelve *Drosophila*	Caspases (CASPs)	Bryant et al. [43]
Culex quinquefasciatus	*Anopheles gambiae* *Aedes aegypti* *Drosophila melanogaster*	Rec, Sig, Mod, Eff	Bartholomay et al. [9]
Acyrthosiphon pisum	*Drosophila melanogaster* *Anopheles gambiae* *Tribolium castaneum* *Apis mellifera* *Pediculus humanus*	Rec, Sig, Mod, Eff	Gerardo et al. [10]
Anopheles gambiae	*Culex quinquefasciatus* *Aedes aegypti*	Mosquito leucine-rich repeat immune proteins (LRIMs)	Waterhouse et al. [44]
Bombyx mori	*Drosophila melanogaster* *Anopheles gambiae* *Aedes aegypti* *Apis mellifera* *Tribolium castaneum*	Serine protease inhibitors (SRPNs)	Zou et al. [45]
Bombyx mori	*Drosophila melanogaster* *Anopheles gambiae* *Apis mellifera* *Tribolium castaneum*	Rec, Sig, Mod, Eff	Tanaka et al. [11]
Drosophila melanogaster	Eleven other *Drosophila*	Rec, Sig, Mod, Eff	Sackton et al. [12]
Aedes aegypti	*Anopheles gambiae* *Culex quinquefasciatus* *Drosophila melanogaster*	Rec, Sig, Mod, Eff	Waterhouse et al. [13]
Tribolium castaneum	*Drosophila melanogaster* *Anopheles gambiae* *Apis mellifera*	Rec, Sig, Mod, Eff	Zou et al. [14]
Apis mellifera	*Drosophila melanogaster* *Anopheles gambiae*	Rec, Sig, Mod, Eff	Evans et al. [15]
Anopheles gambiae	*Drosophila melanogaster*	Rec, Sig, Mod, Eff	Christophides et al. [16]

Gene categories: *Rec* recognition, *Sig* signaling, *Mod* modulation, *Eff* effectors they should be treated as starting points for compiling your own sets of reference sequences. By subsequently curating these initial sets, you will be able to match them with the most up-to-date information, both with respect to the latest genome assembly versions and their corresponding gene sets, as well as to incorporate new discoveries or refinements described in the current literature. One advantage of having selected reference species with publicly browsable genomic resources is that it allows you to perform online queries with gene identifiers or names from the literature in addition to the sequence homology searches described below. Typically, the collected reference sequences will be the translated protein products of each transcript comprising each gene (*see* **Note 3**), stored in plain-text files in FASTA format. When alternative splicing produces protein products that differ substantially (e.g., a single PGRP gene that can encode one, two, or three distinct PGRP domains), it is important to collect all predicted transcripts. This will allow you to assess whether the target species genome also encodes equivalent transcripts and whether gains or losses of alternative transcripts have occurred.

2.1.2 Searching Gene Sets for Candidate Immunity Genes

1. The purpose of compiling a comprehensive and up-to-date set of reference sequences is to then use these as query sequences to search the gene set of the target species being investigated. Your searches should start with a global comparison (*see* **Note 4**) of the compiled sets of reference sequences against the target species' gene set. Use the BLASTp option of the Basic Local Alignment Search Tool (BLAST) suite [46] to identify the most significant matches (i.e., the highest bitscores and the lowest expectation values) to the reference protein sequences in the predicted target proteome (the translations of the predicted gene set). The National Center for Biotechnology Information (NCBI) BLAST+ user manual (https://www.ncbi.nlm.nih.gov/books/NBK279690) provides detailed installation and usage instructions, and example commands (in `monospace` type following `$` symbols) for the required steps are provided here with default parameters:

   Format the protein sequences from your gene set into a searchable database:

```
$ makeblastdb -in geneset_proteins.fasta
-dbtype prot -out proteinsDB
```

   Search your compiled reference protein sequences against the gene set:

```
$ blastp -query reference_proteins.fasta -db
proteinsDB -out referencesVSgeneset.txt
```

   Produce tabular results of searching your compiled reference protein sequences against the gene set:

```
$ blastp -query reference_proteins.fasta -db
proteinsDB -outfmt 6 -out referencesVSgen-
esetTAB.txt
```

The BLASTp search will provide ranked lists of putative homologs of each query sequence from the reference proteins, thereby identifying the predicted proteins encoded in the target genome that most closely resemble the reference sets of immunity proteins. You should next run reciprocal BLASTp searches using the top-scoring proteins from the target species as queries against the complete protein set from the reference species. Your reciprocal searches should return the original query protein as the top-scoring match, especially in the case of proteins encoded by immunity genes that are generally maintained across most species as single-copy orthologs (but *see* **Note 3**). In contrast, for multi-copy gene families, several proteins encoded by members of the gene family in the reference genome may be among the best-scoring matches. These reciprocal sequence homology searches will provide support for the lists of putative immunity genes, but you will need to perform downstream phylogenetic analyses (*see* Subheading 2.1.3 **step 6** below) in order to confirm single-copy orthologs and resolve the relationships among members of multi-copy gene families.

2. The next step is to complement the global protein-protein homology searches of gene set with protein-domain-level searches. Run InterProScan [47] on the proteins from the target species' gene set and the reference protein sequences to obtain detailed domain-level annotations of all protein sequences with significant matches to profiles from the InterPro member databases [48]. Next, use the InterPro domains that characterize each of the different immune gene families or pathway members (Table 2) to identify genes from the target species that encode proteins with significant matches to these domains (*see* **Note 5**). For example, serine protease inhibitors (serpins or SRPNs) are recognized by the "Serpin superfamily" (IPR036186) or "Serpin family" (IPR000215) profiles, or related profiles such as "Serpin, conserved site" (IPR023795) or "Serpin domain" (IPR023796). Exercise caution when the characteristic domains are promiscuous, meaning when they are also present in gene families unrelated to immunity, or when two or more distinct domains characterize a particular immune gene family. For example, Toll-like receptors (TLRs or TOLLs) contain "Leucine-rich repeat" domains, but these are also found in many other types of proteins so their presence is not, on its own, diagnostic of TOLLs. Instead, TOLLs are more specifically characterized by several "Leucine-rich repeat" domains followed by a "Toll/interleukin-1 receptor homology (TIR) domain." The European

Bioinformatics Institute provides detailed InterProScan installation and usage instructions (https://www.ebi.ac.uk/interpro/interproscan.html); the example here uses profiles from the Pfam database:

Scan the gene set protein sequences and compiled sets of reference sequences for matches to InterPro domains:

```
$ ./interproscan.sh -appl Pfam -i geneset_proteins.fasta -f tsv -iprlookup
```

```
$ ./interproscan.sh -appl Pfam -i reference_proteins.fasta -f tsv -iprlookup
```

3. A third approach to searching the target species' gene set for candidate immunity genes is to use profiles built from the reference sequences. First, align each set of orthologous or homologous reference immunity protein sequences collected from several reference species using tools such as PRANK [49] or MAFFT [50]. Next, convert the resulting multiple protein sequence alignments into sequence profiles using HMMER [51]. The HMMER suite of tools can then be used to search the profiles against the target species' gene set. Here we present some examples of the commands that need to be run, but please see the user guides and installation instructions for the alignment tools and HMMER for full details. The input proteins in FASTA format should consist of orthologs or homologs from each of the reference species. Specifically, each FASTA file should contain only proteins encoded by homologs of a single gene or conserved gene family, and the entire analysis should be repeated for each gene or gene family in the study.

Multiple protein sequence alignment example using PRANK:

```
$ prank -d input_proteinset1.fasta -o aligned_proteinset1.aln
```

Multiple protein sequence alignment example using MAFFT:

```
$ mafft input_proteinset1.fasta > aligned_proteinset1.aln
```

```
Convert a multiple protein sequence alignment to a profile using HMMER:
```

```
$ hmmbuild proteinset1.hmm aligned_proteinset1.aln
```

Combine all your profiles into a single profile library (here just three sets shown):

```
$ cat proteinset1.hmm proteinset2.hmm proteinset3.hmm >profile_library
```

Compress and index the library of profiles:

```
$ hmmpressprofile_library
```

Search the library of profiles against the target species' gene set using HMMER:

```
$ hmmscanprofile_librarygeneset_proteins.
fasta
```

2.1.3 Curating Candidate Immune-Related Genes

1. Your global protein sequence and profile searches and protein-domain searches will result in lists of candidate immune-related genes from the target species. With good supporting data, especially from transcriptomics (as described below in Subheading 2.2), automated prediction pipelines applied to well-assembled genomes generally produce gene sets with a high coverage of the true gene content [52–54]. The task nevertheless remains challenging, and accurate predictions at the detailed level of gene intron-exon structures can be difficult to achieve even with extensive supporting data. Manual curation aims to verify that the automatically predicted gene models identified through your sequence and domain searches are in agreement with the available supporting evidence. You may undertake the curation process with a small team or you may bring together several groups of researchers and/or students (e.g., [55–57]) to examine your lists of candidate immunity genes. For a small team, the curation process may focus on quality control and targeted appraisal of specific genes of interest. For example, quality control of seemingly anomalous results can confirm true novelties, such as the multi-PGRP-domain PGRP proteins encoded in the banded demoiselle genome [58]. For a larger research community, the aims may be broader and may include taking advantage of researchers' expertise to build a rich knowledge base for the target species. The tools and approaches described here are useful for both small- and large-scale curation efforts.
2. Several computational resources need to be set up so that the genomic data from the target species can be easily queried by users with little or no bioinformatics expertise. You can achieve a local setup of the necessary resources with relatively modest computational equipment and the installation of several freely available bioinformatics packages and software. The key components should include a genome browser and a sequence search interface. A particularly useful platform that allows for sequence-based database searching is the combination of the JBrowse genome viewer [59] with the Apollo annotation feature editor plug-in [60] and SequenceServer [61]. Software installation is beyond the scope of this chapter but is described in detail in the respective setup and user guides. These resources will provide you with a user-friendly environment to interrogate the genomics data without requiring experience with running command-line bioinformatics tools. They also offer the flexibility to search gene-by-gene for specific genes of

interest, to search using sequences from species or genes that were not included in the compiled sets of reference sequences, or to use sequences from the target species to search for within-species homologs.

3. A tBLASTn search of the reference immunity sequences against the target species' genome assembly will enable visualization of genomic loci with homology to the reference proteins. tBLASTn uses the provided reference protein sequences to search the six-frame translations of the genome assembly nucleotides and is more sensitive than nucleotide-nucleotide searches. The tBLASTn results are useful because the automated pipeline used to predict gene models in the target species may have missed or misannotated some genes or exons, meaning that they would be impossible or difficult to identify from searching only the predicted gene set. You should produce tabular format outputs of the tBLASTn searches because these can be loaded as data tracks for visualization within a genome browser after converting them into general feature format (GFF) output files (*see* **Note 4**). The following commands illustrate how this can be achieved:

   Format your genome assembly into a searchable database:

```
$ makeblastdb -in genome_assembly.fasta
-dbtype nucl -out assemblyDB
```

   Produce tabular results of searching your compiled reference protein sequences against the genome assembly:

```
$ tblastn -query reference_proteins.fasta
-db assemblyDB -outfmt 6 -out referencesVS-
assemblyTAB.txt
```

4. The locations of the best hits define genomic loci that likely encode orthologs or homologs of the reference sequences. Visualizing these using a genome browser enables you to assess how much of the reference sequence aligns to the target assembly and how well these alignments match up to the predicted gene model (*see* **Note 6**). Complementary supporting evidence comes from transcriptomics data in the form of RNA sequencing (RNA-seq) reads from samples prepared from your target species. The RNA-seq reads may derive from your own infection experiments (*see* Subheading 2.2 below), but if other datasets are available, then it is advisable to also include these as additional supporting data. You will need to align the reads to the genome assembly in order to visualize them in a genome browser, typically as both stacked individual read alignments and read coverage plots (*see* **Note 4**). Several bioinformatics tools are able to align reads to an assembly (e.g., HISAT2 [62] or STAR [63]), and coverage plots can be built using bamCoverage from the deepTools suite [64]. Here we

present some examples of the commands that need to be run, but please see the user guides for full details.

Build an index of your genome assembly and then align fastq format RNA-seq reads using HISAT2:

```
$ hisat2-build genome_assembly.fastaindex_
name
```

```
$ hisat2 -x index_name -1 sample_1.fastq -2
sample_2.fastq -S hisat2-mapped.sam
```

Build an index of your genome assembly and then align fastq RNA-seq reads to your assembly using STAR:

```
$ STAR --runModegenomeGenerate --genomeDir
star-index --genomeFastaFiles genome_assem-
bly.fasta
```

```
$ STAR --genomeDir star-index --readFilesIn
sample_1.fastq sample_2.fastq --outSAMtype
BAM SortedByCoordinate
```

Produce an RNA-seq read coverage file using bamCoverage:

```
$ bamCoverage -b Aligned.sortedByCoord.out.
bam -o rnaseq-coverage.bw
```

5. With the necessary resources in place, the next step is to examine the genomic locus encoding each candidate immunity gene in order to establish whether the predicted model is well supported (*see* **Note 7**). Well-supported models generally show RNA-seq coverage and spliced RNA-seq read alignments that match the intron-exon structure of the entire model and tBLASTn alignments for most of the model. Typical minor edits to improve the models include altering the intron-exon boundaries to match the aligned RNA-seq reads, removing non-supported exons (i.e., predicted exons that have no tBLASTn alignments and no aligned RNA-seq reads), or adding exons missed by the automated prediction pipeline (i.e., regions with tBLASTn alignments and/or aligned RNA-seq reads where no exon was predicted). For example, Fig. 2 shows how editing an incorrectly predicted intron-exon boundary to match the supporting RNA-seq read alignments produces a full-length gene model for *Dicer-2*. More substantial edits include the merging of two or more neighboring predicted gene models that in fact encode a single gene, or the splitting of gene models where the automated gene prediction has incorrectly fused neighboring genes. Automated gene predictors are prone to such erroneous fusing of neighboring genes when the genes are homologous or have arisen from tandem gene duplication events. Thus it is worth paying particular attention to the gene model predictions of members of

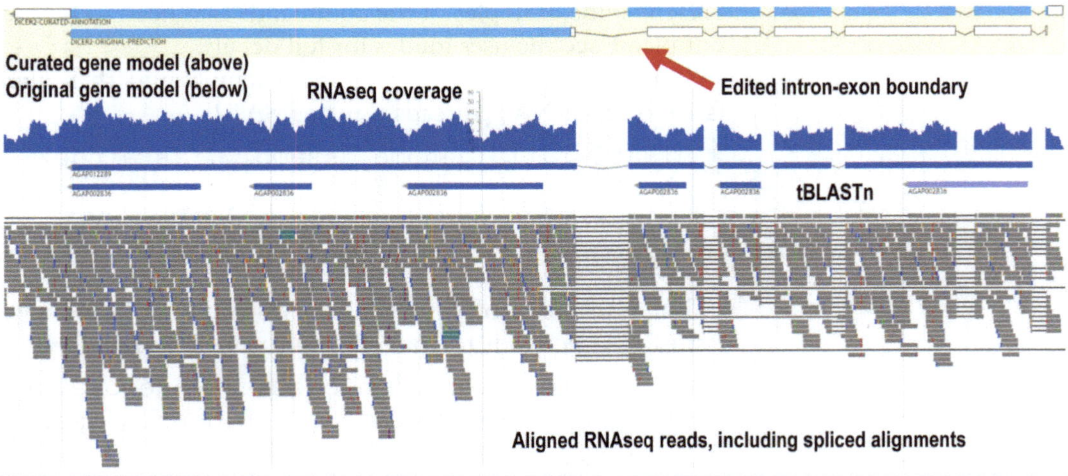

Fig. 2 Example of how manual curation can improve automatically predicted gene models. The top panel shows the curated gene model and the original prediction of the *Dicer-2* gene on the reverse strand (i.e., the 5′ start is on the right and the 3′ end is on the left of the figure) from a mosquito genome. Exons are shown as rectangles connected with lines indicating introns, with predicted coding sequence (CDS) regions in light blue and predicted untranslated regions (UTRs) shown in white. RNA-seq read coverage is presented below the gene models in dark blue, clearly showing where reads from the mature messenger RNA align to the genome. Below that are alignments from tBLASTn searches with the Dicer-2 protein (AGAP012289) and the Dicer-1 protein (AGAP002836) from *Anopheles gambiae* (the reference immune protein sequences). The lower panel shows the alignments of individual RNA-seq reads to this locus (in dark gray, with colors indicating mismatches between the reads and the reference genome assembly), with reads that map across potential splice junctions connected with black lines. Editing just one intron-exon boundary to match the supporting RNA-seq and tBLASTn evidence (shown with the red arrow) corrects the gene model. The first six exons were incorrectly predicted to form a multi-exon 5′ UTR (all white rectangles) in the original gene model. In the curated gene model, all six exons now form part of the CDS (i.e., the regions that will be translated into protein), with just a short 5′ UTR at the start of the first exon. The translation of the curated gene model now encodes a full-length Dicer-2 protein

multi-copy gene families. In addition, it is often challenging for automated pipelines to correctly predict two or more alternative transcripts from the same gene, so manual editing may be required to distinguish the individual transcripts based on the available supporting data.

6. One reason for checking and correcting the candidate immune-related gene models is to facilitate subsequent phylogenetic analysis of immune genes or gene families of particular interest, including where putative duplications/expansions have been noted from the initial searches. Molecular phylogenetic analysis aims to reconstruct the evolutionary histories of sets of homologous sequences. Conceptually, this is achieved by contrasting the species phylogeny with the inferred gene trees to enable the confident assignment of orthologous relations [65]. In practice there are many different methodological approaches and

bioinformatics tools designed for preparing and analyzing the sequence data required for phylogenetic tree construction, the discussion of which is beyond the scope of this chapter. One suite of such tools that is particularly user-friendly for novices is the Molecular Evolutionary Genetics Analysis (MEGA) software [66]. In the context of characterizing your sets of newly identified putative immune-related genes, the phylogenetic analyses will allow you to (1) confirm or refine orthologous relations suggested by your reciprocal sequence homology searches and (2) place putative gene duplications or losses in their appropriate evolutionary contexts.

2.2 Identification of Infection-Responsive Genes

While searching based on sequence homology is a valuable approach to identify canonical immune genes in new species, some immunologically important genes may be novel to the target species or otherwise difficult to identify from sequence data. In many cases, however, expression of these genes is responsive to infection [21]. These can include both genes that are directly involved in immune defense and also genes that are regulated as a consequence of infection. Using RNA sequencing (RNA-seq), it is possible to obtain a direct readout of the transcriptional response to infection.

There are a number of important experimental design issues to consider before embarking on RNA-seq-based identification of immune-responsive genes [67]. Two key requirements must be met for a successful experiment. First, in order for the protocol outlined below to be successful, a mostly complete draft genome with a high-quality gene set must exist for the target insect. While it is possible to use RNA-seq data to build a de novo transcriptome [68, 69] (and *see* Chapter 2 of this book) or to aid gene prediction for a draft genome without a gene set [62, 70], this is beyond the scope of this chapter and we do not recommend it unless there is no alternative. Second, it must be possible to experimentally infect the target insect in the laboratory. Ideally, the insect can be maintained for several generations under controlled conditions to eliminate effects of previous exposure to pathogenic challenges or other stimuli that could modulate the immune response.

The simplest experimental design to identify genes that are transcriptionally responsive to infection would include just a single control condition (either naive, untreated insects or sterilely wounded insects) and a single experimental condition at some time post-infection with the desired infectious challenge. More complex designs could include multiple controls, multiple pathogenic agents, and/or multiple time points. As a general rule of thumb, a minimum of three biological replicates should be included for each experimental treatment and control, although additional replicates will increase statistical power [71–74]. If the target insect is so small that sufficient RNA is hard to obtain from a single insect, pools of genetically similar (or ideally identical) individuals can be used, but this does not eliminate the need for multiple biological replicates of the experiment.

2.2.1 Artificial Infections for RNA-Seq Analysis

Insects mount different immune responses to different types of infectious challenge (e.g., bacterial, fungal, viral, protozoan, nematode, etc.), and different challenges will therefore elicit different transcriptional responses. Injection with bacteria or bacterial cell wall and membrane components is often used as a generic immune stimulus for identification of genes that are transcriptionally responsive to infection [19, 20]. Here, we detail a protocol for infection of a small insect like *Drosophila* or a mosquito with a live bacterium. The protocol is demonstrated visually in [75] and can be modified for larger insects or for other infectious agents. The experimenter should choose the most appropriate challenge for the system being queried and modify delivery of the challenge accordingly.

1. In order to minimize experimental noise, all insects should be reared in the laboratory without exposure to pathogens prior to the experiment. This will allow optimal comparison of the expression profiles of infected insects to unchallenged controls. Biological replicates should be collected for both challenged and unchallenged insects (*see* **Note 8**). For small insects or small tissue samples taken from larger insects, the material from multiple individuals can be pooled within each biological replicate. Using co-reared insects that are the same age and sex will minimize experimental noise, although in some cases it may be of interest to make comparisons across life stages, sexes, or rearing conditions (*see* **Note 9**).
2. Culture the infectious agent and prepare it for infection. In the case of bacterial challenge, infection may be delivered with a single bacterium or a mixture of different bacteria, and the bacteria may be either alive or killed by incubation at 60 °C for 30 min (*see* **Note 10**).
3. Challenge the insects in the infection treatment. Bacteria, planktonic fungi, and viruses can be injected into insects with a microcapillary needle. Live bacteria may also be introduced with a septic pinprick (demonstrated in detail in [75]) (*see* **Note 11**). Other challenges, such as infection with filamentous fungi (e.g., [76]) or eukaryotic parasites (e.g., [77]), require different methods.
4. Collect the insects at the prescribed time point post-infection (*see* **Note 12**). RNA may be isolated immediately or the insects may be flash-frozen in liquid nitrogen and stored at −80 °C until RNA extraction is to be performed. If RNA will be performed using a TRIzol (Invitrogen) extraction, the insects or insect tissue may be stored at −80 °C in TRIzol.

5. Isolate high-quality RNA from the infected and control insects. There are a variety of protocols and commercial kits available for RNA isolation, and any of these should work well for RNA sequencing. Isolations using TRIzol reagent (Invitrogen) are reliable and inexpensive. A thorough protocol for RNA isolation using TRIzol is outlined in Chapter 2 of this volume. Consult with the facility that will perform your RNA sequencing to see whether they have preferences or recommendations as to which RNA isolation procedure should be employed.
6. Perform the RNA sequencing (RNA-seq) on your infected and control insect material. In most circumstances, we recommend that inexperienced practitioners outsource library preparation and sequencing to a core facility or commercial provider. The library preparation is highly technical and labor intensive, and the technology changes quickly. Unless a very large number of libraries are going to be generated, the cost savings associated with doing the preparation yourself are generally not worth the effort or the risk of failed reactions. Therefore, if possible, use a facility that will accept RNA shipped on dry ice and that prepares their libraries and performs sequencing in-house. The optimal read length and depth of sequencing will depend on project budget and a variety of other factors that will vary among projects. For the analysis described below, we recommend a minimum of ten million fragments sequenced per replicate, using at least 40-bp paired-end reads. Increasing read depth to 20–30 million fragments per replicate can be beneficial if project scope and funding allow (*see* **Note 13**), and increasing read length to 75 bp will decrease the number of reads that map ambiguously to multiple locations in the genome (e.g., reads from members of closely related gene families).

2.2.2 Performing Differential Expression Analysis

1. The first step in differential expression analysis is using a read alignment or pseudoalignment (*see* **Note 14**) to estimate expression of each transcript or gene (*see* **Note 15**). Here we present one option for this, but there are many alternative choices (*see* **Note 16**). The protocol here assumes you have paired-end sequencing reads from your core facility or commercial provider, in fastq format. We describe optional quality control and trimming steps in **Note 17**. A workflow of the steps required to perform differential expression analysis is presented in Fig. 3. In the following steps, command lines are given with variables (file names, species, and sample identifiers) that will need to be changed for each experiment in curly braces {}. Commands are given in `monospace type`.
2. This protocol uses commands from the kallisto program [78] (https://pachterlab.github.io/kallisto/) and should run in

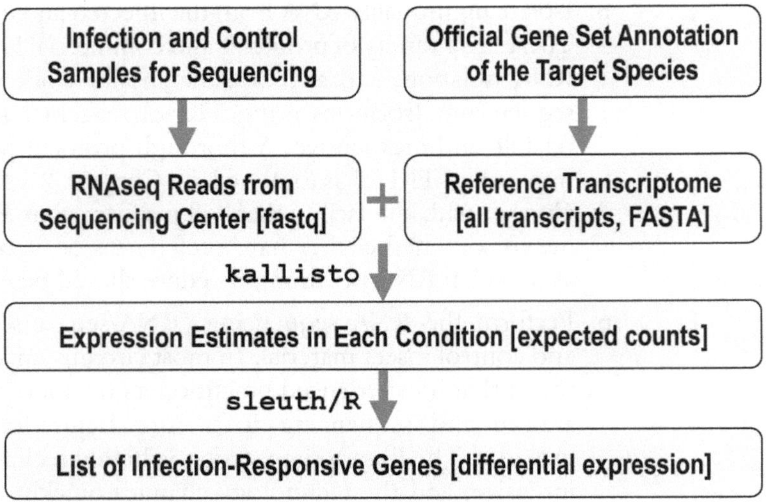

Fig. 3 Workflow of steps required for immune transcriptome analysis. Immune transcriptome analysis can proceed once the RNA-seq reads (in fastq format) from all the infection and control samples have been obtained. The analysis also requires the complete set of transcripts from the gene set annotation of the target species, which may also contain updated gene model annotations based on manual curation described in Subheading 2.1.3 of this chapter. In the kallisto documentation, this complete set of transcripts is referred to as the "reference transcriptome" to which the RNA-seq reads will be mapped. RNA-seq reads (possibly after pre-processing; *see* **Note 17**) are mapped to transcripts by kallisto using a pseudoalignment step that then allows for the quantification of transcript abundances from each condition to determine expression levels of each gene and isoform. Finally, differential expression of genes and isoforms among conditions is modeled using sleuth/R to define sets of infection-responsive genes

less than an hour per sample on a typical laptop computer. Software installation is beyond the scope of this chapter but is described in detail here: https://pachterlab.github.io/kallisto/download. The first step in using kallisto is to prepare the index. Indexing takes a plain-text FASTA file containing the nucleotide sequences of all transcripts from the gene set of a given genome and converts it into a format that allows for subsequent rapid pseudoalignment of the RNA-seq reads to the transcripts. The complete set of transcripts from the gene set to be analyzed is referred to in the kallisto documentation as the "reference transcriptome" to which the RNA-seq reads will be mapped. For your target species, you should obtain the FASTA file of transcripts from the official gene set provided by public databases (e.g., Ensembl, FlyBase, NCBI, VectorBase). If the data are only available in-house, then use the FASTA file of transcripts resulting from the full genome annotation pipeline.

3. Prepare a reference transcriptome index for kallisto. First, make a working directory and copy the transcriptome FASTA file to it. You can then index this file and proceed to quantify transcript abundances. You will obtain a {SAMP}_out direc-

tory for each sample/replicate you generated, which can be used with sleuth (or other tools) as described below to estimate differentially expressed transcripts and genes per condition.

In the working directory and with kallisto installed:

```
$ kallisto index -i {INDEX_NAME}.idx
{TRANSCRIPTOME}.fasta
```

Quantify abundance of transcripts in each sample, where {SAMP} is the fastq base name for a particular replicate/condition:

```
$ kallisto quant -i {INDEX_NAME}.idx -o
{SAMP}_out -b 100 {SAMP}_R1.fastq.gz
{SAMP}_R2.fastq.gz
```

4. There are many toolkits for detecting genes with differential expression between conditions. Here we present protocols for using sleuth [79], but discuss alternatives in **Note 18**. Note that sleuth requires the technical bootstraps generated by kallisto for full functionality, and thus we only recommend this protocol to be used with data analyzed first by kallisto.

   Open R and ensure that the sleuth package is installed, as well as tidyverse which is used for some data manipulation tasks (*see* **Note 19**):

```
$ library(sleuth)
$ library(tidyverse)
```

   Set the path to your kallisto output files:

```
$ kall_path <- {PATH/TO/FILES}
```

   Get sample identifiers from names of kallisto runs:

```
$ sample_id <- dir(file.path(kall_path))
```

   Get the directories where the kallisto runs are saved:

```
$ kal_dirs <- data.frame(sample_id = sample_
id, path = file.path(kall_path, sample_id))
```

   Load the table that associates sample identifiers with treatments and add file paths. You will need to create this yourself (*see* **Note 20**):

```
$ s2c<-read_table("{PATH/TO/TABLE}")%>%
full_join(kal_dirs, by=c("sample_id" =
"sample_id")
```

   Load gene to transcript map (*see* **Note 21**):

```
$ t2g<-read_table("{T2G_FILE"})
Run sleuth prep; note this aggregates tran-
script-level counts into gene-level counts:
$ so<-sleuth_prep(s2c, extra_bootstrap_
summary=TRUE, read_bootstrap_tpm=TRUE,
```

```
target_mapping = t2g, aggregation_column =
'gene_id')
```

Fit a sleuth model (*see* **Note 22**):

```
$ so<-sleuth_fit(so, ~treatment, 'full')
$ so<-sleuth_wt(so, "inf", which_model =
"full")
```

Output results:

```
de_genes<- sleuth_results(so, test="inf")
```

Note that there are many quality control and plotting options available in sleuth, which can be explored using the built-in Shiny Server. To launch run:

```
$ sleuth_live(so)
```

3 Notes

1. In addition to the references presented in the introduction, literature reviews that focus on different pathways or responses can provide additional details as to the expected structure and function of immune system components (e.g., on antiviral immunity [80], or the IMD [81], JAK/STAT [82], or Toll [83] pathways). While studies of the *Drosophila* immune system provide a rich knowledge base for understanding insect immunity, this model should be considered as a sample of the full-spectrum immunity in insects. Experimental examination of immune responses in other insects has revealed many features that are widespread, such as melanization reactions and presence of the principal immune signaling pathways. However, they have also identified many lineage-specific features that differ greatly from observations to date in flies. For example, adult *Drosophila* have very few circulating hemocytes (blood cells) [84] so the relative importance of cellular immunity is probably underestimated in *Drosophila* relative to other insects. With the great diversity of insect species (over 500 million years of evolution) and the variety of pathogens they encounter in their various ecological niches, such differences are to be expected.
2. Immune-related genes of the canonical repertoire in fact comprise many genes that may not have direct experimental evidence supporting their roles in immunity. It is also important to note that many genes and pathways have pleiotropic functions, meaning a single gene can produce a protein that is involved in different biological processes, so being classified as a canonical immunity gene does not preclude involvement in other processes. Similarly, the sub-classification of genes into recognition, signal transduction, modulation, or

defense/effector phases is a useful framework, but it does not necessarily exclude the possibility of the protein being involved in other processes.

3. For gene models with alternative transcripts, it is advisable to collect the sequences for each transcript that produces a distinct protein product through alternative splicing, because (1) annotation prediction of alternative transcripts by automated pipelines is particularly challenging so having a reference set of possible transcripts will help to build accurate gene models during curation and (2) being able to select equivalent transcripts will make downstream phylogenetic analyses more robust and, in the case of alternatively spliced protein domains, will allow for domain-based analyses. It should also be noted that sequence homology searches with the different protein products of alternative transcripts may obscure truly reciprocal best matches at the level of the gene. These can generally be resolved by examining the genomic loci to determine equivalence at the transcript level.
4. Performing global searches of all the compiled sets of reference protein sequences against the proteins from the gene set will require running some bioinformatics sequence analysis tools. Working with colleagues who have experience running such analyses will allow novice team members to learn these key skills. Installing the required software and setting up the resources to run a local genome browser and sequence search interface can be achieved with a range of freely available bioinformatics tools. Aligning RNA-seq reads to the genome assembly and producing tracks for visualization in a genome browser will greatly facilitate the process of manually curating the candidate immune-related genes. Providing detailed instructions for installing and running these tools is beyond the scope of this chapter. Instead, team members should be able to relatively easily set up these necessary resources following instructions in the references and links provided herein. These tools will greatly facilitate both the gene identification and curation steps, e.g., being able to visualize the genomic locations of the sequences that produce significant matches to the reference protein sequences (using the tabular tBLASTn results) in order to find genes that may have been missed by the automated gene prediction pipeline as well as highlighting possible errors in the predicted gene models that need to be corrected during manual curation.
5. Examining the results from running InterProScan on the compiled sets of reference proteins will provide an up-to-date summary of which proteins encoded in the target genome contain domains that are characteristic of members of the canonical immune gene repertoire. It is important to note that InterPro

entry types range from general to specific: homologous superfamily, protein family, domain, repeat, or site. Thus the more general entry types may recognize a much broader set of proteins than the immune genes of interest. For example, the prophenoloxidases (PPOs) are recognized by the "Hemocyanin/hexamerin" family (IPR013788) profile, which also recognizes insect hexamerins (storage proteins).

6. The alignments that define significant matches between the reference protein sequences and the target assembly are not expected to correspond perfectly to the predicted gene model in the target species. Evolutionary divergence between the reference and target species means that only the relatively well-conserved regions of most proteins will produce confident alignments. Highly diverged regions, regions of low-complexity sequence, and short exons may produce no significant hits and therefore could appear as non-supported parts of the gene model. In addition, the alignment boundaries are unlikely to match exactly the intron-exon boundaries of the gene model since tBLASTn searches do not take putative splice sites into account. Thus, the homology searches serve to identify the most likely genomic loci encoding genes of interest and they provide support for the predicted gene model, but differences between the alignment coordinates and the gene model are to be expected.
7. Detailed practical guidelines for performing manual curation of predicted gene models and assessing the supporting evidence using the Apollo online collaborative genomic annotation editor are provided in the documentation and user guide materials (http://genomearchitect.github.io). Additional training materials include several webinars available through YouTube, e.g., from the Bioinformatics Platform for Agroecosystem Arthropods https://www.youtube.com/watch?v=BMeSwdKiO_E or from the European Molecular Biology Laboratory Australia Bioinformatics Resource https://www.youtube.com/watch?v=Wec7ZlXykQc.
8. The simplest possible experimental design is a single control (three replicates of either untreated insects or sterilely wounded insects) compared to three replicates of infected insects assayed at a single time point post-infection. More complicated experiments might include a time series after infection to capture transcriptional dynamics in response to infection. Depending on the goals and scope of the project, a variety of options are feasible. More complex designs (e.g., those with more than a single control and a single infected treatment) will require more complicated analysis.
9. Exact age of insects will depend substantially on the species and goals of the project (e.g., comparisons across life stages or

sexes may be of interest). In general, to minimize uncontrolled noise, ensuring that the experimental insects are of roughly the same age and the same sex is standard practice. The number of individual insects depends on size and the amount of RNA that can be obtained from single individuals. Your sequencing provider can tell you how much starting material is necessary for library preparation, which provides a starting point for the infection experimental design.

10. Challenge with a single bacterial strain will give a clean measurement of the transcriptional response to that bacterium, whereas challenge with a pool of bacterial species (e.g., including both Gram-negative and Gram-positive) will reveal a broader spectrum of responses but will not allow determination of which genes are responding to which microbe. Live bacterial infection will stimulate transcriptional responses to both the presence of bacteria (e.g., immune stimulation by peptidoglycan) as well as responses to pathogenic damage caused by infection, which can also be a strong trigger of immune responses [85]. The ideal bacterial concentration is one that is sufficient to induce a strong immune response without causing substantial mortality so that immune responses do not become conflated with transcriptional signatures of death. In most cases pilot experiments using different concentrations and measuring mortality over time will be necessary to calibrate the proper dosage. Challenge with dead bacteria or purified bacterial components eliminates concerns about host mortality and often is sufficient for stimulating a robust response [25]. It should be noted that some pathogens are capable of suppressing host responses (e.g., [86]), so heat-killing these prior to infection may yield a stronger response. Pathogens such as viruses, nematodes, and protozoa generally need to be alive in order to infect so these should not be heat-killed unless required by the specific objectives of the experiment. A standard method for culturing bacteria prior to infecting *D. melanogaster* is shown visually in [75].
11. Delivering infection by septic pinprick is less quantitatively controlled than performing injections with a microcapillary needle but also requires less equipment and technical proficiency. For many experimental designs, especially those using a mixed pool of bacteria to elicit a broad-spectrum immune response, precise quantification of the challenges is probably unnecessary. It should be noted, however, that septic pinprick delivers fairly low infection dose that may not be sufficient to stimulate a robust response in large insects such as large caterpillars and beetles. For these insects, microcapillary injection may be required.

12. The time after infection at which to measure expression is an important decision. Bacterial infections elicit a rapid response in insects, and sampling at 8–12 h post-infection is common and experimentally convenient (allowing infection in the morning and freezing of infected insects in the evening, or infections in the evening and freezing the following morning) [7, 87, 88]. However, transcriptional dynamics vary depending on the pathogenic agent and other experimental variables [25, 89]. Therefore it is advisable to perform preliminary experiments before collecting samples for sequencing to be able to select the most appropriate conditions and time points. These pilot studies could involve low-coverage RNA-seq from a single sample across multiple time points or could involve quantitative PCR of candidate immune effectors, such as antimicrobial peptides, that provide reliable readouts of immune system activation.
13. In general, power to detect differential expression scales more with replicate number than with reads per sample [71]. So for a fixed amount of sequencing, there is more experimental gain in sequencing a greater number of replicates to individually lower depth than sequencing fewer replicates to higher depth. However, given a fixed number of replicates, increasing depth will also increase resolution and power up to a point. Sequencing depth can be adjusted to the scope of the project and available budget.
14. There are two approaches to determining which transcript a read arises from. The traditional approach uses standard read alignment metrics to map a particular read to a genome (or transcriptome) sequence and then uses the mapping position to determine the transcript. There are many programs that can perform this alignment procedure, as recent benchmarking studies show [90]. The pseudoalignment approach instead uses representations of transcripts and reads to find a fast match; this has the benefit of greatly increased speed and computational efficiency, at no cost to accuracy [91].
15. For the purposes of identifying genes regulated by infection, aggregating results to gene-level summaries (in which expression values are aggregated across all alternative isoforms of a gene) is often the most desirable outcome. There is some debate about the best way to do this, e.g., [92]; we have presented one option but there are alternatives such as those described in the discussion here: https://pachterlab.github.io/sleuth/walkthroughs. In addition, when evaluating alternative splicing and related questions, it is essential to estimate transcript-level differential expression instead of gene-level differential expression.
16. We present a method using kallisto [78] to generate expression estimates for use in downstream pipelines, but there are

several alternatives, including salmon, which also uses pseudoalignment [93]; RSEM, which uses full alignment [94]; and others. kallisto has the considerable advantage of low compute requirements, meaning a typical experiment can be analyzed on a laptop computer without the need for dedicated computing clusters.

17. Trimming low-quality reads generally is not necessary for RNA-seq differential expression analysis, although removing adaptors can be useful if your reads have substantial adaptor contamination. There are a number of tools for doing this, including Trimmomatic [95] and NGmerge [96].
18. There are a wide variety of R packages that can fit differential expression models to RNA-seq data, including DESeq2 [97], limma-voom [98], and edgeR [99]. We focus on sleuth here, as it is designed to work with the output of kallisto, but all of the listed tools perform well.
19. For most packages, including tidyverse and dependencies (but not sleuth), it should be possible to install them using the `install.packages("{PACKAGE NAME}")` command. See the tidyverse documentation and the sleuth documentation for additional details.
20. Sleuth requires a table that has sample_id as one column and the treatment (e.g., infected, control) as the second column, in order to match samples to conditions. This can be prepared in Excel or similar spreadsheet software, saved as a CSV file, and loaded into R.
21. To aggregate transcript-level results into gene-level counts requires a file mapping transcript identifiers to gene identifiers. This should be a text file with two columns, one with transcript identifiers matching the transcripts used in kallisto and the other with gene_id.
22. Sleuth uses two approaches to estimate significance of differential expression: a Wald test, which compares two conditions, and a likelihood ratio test, which can compare arbitrary nested models. In this case, we show how to run a simple Wald test comparing an infected sample and control sample, for a simple experiment with only two conditions. For more complex experiments, a likelihood ratio test may be more useful. See the sleuth manual for details.

Acknowledgments

RMW acknowledges the Swiss National Science Foundation (grants PP00P3_170664 and CRSII5_186397); the Department of Ecology and Evolution, University of Lausanne; and Swiss

Institute of Bioinformatics, Switzerland. BPL acknowledges the Unites States National Institutes of Health (grant R01 AI141385); and the Cornell Institute of Host-Microbe Interactions and Disease, Department of Entomology, Cornell University, Ithaca, New York, USA. TBS acknowledges the Informatics Group, Faculty of Arts and Sciences, Harvard University, Cambridge, Massachusetts, USA.

References

1. Ligoxygakis P (2017) Advances in insect physiology. volume 52, insect immunity. Academic Press, San Diego
2. Buchon N, Silverman N, Cherry S (2014) Immunity in *Drosophila melanogaster*-from microbial recognition to whole-organism physiology. Nat Rev Immunol 14:796–810. https://doi.org/10.1038/nri3763
3. Imler J-L (2014) Overview of *Drosophila immunity*: a historical perspective. Dev Comp Immunol 42:3–15. https://doi.org/10.1016/j.dci.2013.08.018
4. Rolff J, Reynolds SE (2009) Insect infection and immunity: evolution, ecology, and mechanisms. Oxford University Press, New York. https://doi.org/10.1093/acprof:oso/9780199551354.001.0001
5. Ferrandon D, Imler J-L, Hetru C, Hoffmann JA (2007) The *Drosophila* systemic immune response: sensing and signalling during bacterial and fungal infections. Nat Rev Immunol 7:862–874. https://doi.org/10.1038/nri2194
6. Lemaitre B, Hoffmann J (2007) The host defense of *Drosophila melanogaster*. Annu Rev Immunol 25:697–743. https://doi.org/10.1146/annurev.immunol.25.022106.141615
7. Sackton TB, Lazzaro BP, Clark AG, Wittkopp P (2017) Rapid expansion of immune-related gene families in the house fly, *Musca domestica*. Mol Biol Evol 34:857–872. https://doi.org/10.1093/molbev/msw285
8. Barribeau SM, Sadd BM, du Plessis L et al (2015) A depauperate immune repertoire precedes evolution of sociality in bees. Genome Biol 16:83. https://doi.org/10.1186/s13059-015-0628-y
9. Bartholomay LC, Waterhouse RM, Mayhew GF et al (2010) Pathogenomics of *Culex quinquefasciatus* and meta-analysis of infection responses to diverse pathogens. Science 330:88–90. https://doi.org/10.1126/science.1193162
10. Gerardo NM, Altincicek B, Anselme C et al (2010) Immunity and other defenses in pea aphids, *Acyrthosiphon pisum*. Genome Biol 11:R21. https://doi.org/10.1186/gb-2010-11-2-r21
11. Tanaka H, Ishibashi J, Fujita K et al (2008) A genome-wide analysis of genes and gene families involved in innate immunity of *Bombyx mori*. Insect Biochem Mol Biol 38:1087–1110. https://doi.org/10.1016/j.ibmb.2008.09.001
12. Sackton TB, Lazzaro BP, Schlenke TA et al (2007) Dynamic evolution of the innate immune system in *Drosophila*. Nat Genet 39:1461–1468. https://doi.org/10.1038/ng.2007.60
13. Waterhouse RM, Kriventseva EV, Meister S et al (2007) Evolutionary dynamics of immune-related genes and pathways in disease-vector mosquitoes. Science 316:1738–1743. https://doi.org/10.1126/science.1139862
14. Zou Z, Evans JD, Lu Z et al (2007) Comparative genomic analysis of the *Tribolium* immune system. Genome Biol 8:R177. https://doi.org/10.1186/gb-2007-8-8-r177
15. Evans JD, Aronstein K, Chen YP et al (2006) Immune pathways and defence mechanisms in honey bees *Apis mellifera*. Insect Mol Biol 15:645–656. https://doi.org/10.1111/j.1365-2583.2006.00682.x
16. Christophides GK, Zdobnov E, Barillas-Mury C et al (2002) Immunity-related genes and gene families in *Anopheles gambiae*. Science 298:159–165. https://doi.org/10.1126/science.1077136
17. Waterhouse RM, Wyder S, Zdobnov EM (2008) The *Aedes aegypti* genome: a comparative perspective. Insect Mol Biol 17:1–8. https://doi.org/10.1111/j.1365-2583.2008.00772.x
18. Olafson PU, Aksoy S, Attardo GM, et al (2019) Functional genomics of the stable fly, *Stomoxys calcitrans*, reveals mechanisms underlying reproduction, host interactions, and novel targets for pest control. bioRxiv 623009. https://doi.org/10.1101/623009
19. Altincicek B, Vilcinskas A (2007) Analysis of the immune-inducible transcriptome from

microbial stress resistant, rat-tailed maggots of the drone fly *Eristalis tenax*. BMC Genomics 8:326. https://doi.org/10.1186/1471-2164-8-326
20. Sackton TB, Clark AG (2009) Comparative profiling of the transcriptional response to infection in two species of *Drosophila* by short-read cDNA sequencing. BMC Genomics 10:259. https://doi.org/10.1186/1471-2164-10-259
21. Sackton TB (2019) Comparative genomics and transcriptomics of host–pathogen interactions in insects: evolutionary insights and future directions. Curr Opin Insect Sci 31:106–113. https://doi.org/10.1016/J.COIS.2018.12.007
22. Ekengren S, Hultmark D (2001) A family of Turandot-related genes in the humoral stress response of *Drosophila*. Biochem Biophys Res Commun 284:998–1003. https://doi.org/10.1006/bbrc.2001.5067
23. Brun S, Vidal S, Spellman P et al (2006) The MAPKKK Mekkl regulates the expression of Turandot stress genes in response to septic injury in *Drosophila*. Genes Cells 11:397–407. https://doi.org/10.1111/j.1365-2443.2006.00953.x
24. De Gregorio E, Spellman PT, Rubin GM, Lemaitre B (2001) Genome-wide analysis of the *Drosophila* immune response by using oligonucleotide microarrays. Proc Natl Acad Sci 98:12590–12595. https://doi.org/10.1073/pnas.221458698
25. Troha K, Im JH, Revah J et al (2018) Comparative transcriptomics reveals CrebA as a novel regulator of infection tolerance in *D. melanogaster*. PLOS Pathog 14:e1006847. https://doi.org/10.1371/journal.ppat.1006847
26. Hillyer JF (2016) Insect immunology and hematopoiesis. Dev Comp Immunol 58:102–118. https://doi.org/10.1016/j.dci.2015.12.006
27. Bartholomay LC, Michel K (2018) Mosquito immunobiology: the intersection of vector health and vector competence. Annu Rev Entomol 63:145–167. https://doi.org/10.1146/annurev-ento-010715-023530
28. Legeai F, Shigenobu S, Gauthier JP et al (2010) AphidBase: a centralized bioinformatic resource for annotation of the pea aphid genome. Insect Mol Biol 19(Suppl 2):5–12. https://doi.org/10.1111/j.1365-2583.2009.00930.x
29. Kersey PJ, Allen JE, Allot A et al (2018) Ensembl genomes 2018: an integrated omics infrastructure for non-vertebrate species. Nucleic Acids Res 46:D802–D808. https://doi.org/10.1093/nar/gkx1011
30. Thurmond J, Goodman JL, Strelets VB et al (2019) FlyBase 2.0: the next generation. Nucleic Acids Res 47:D759–D765. https://doi.org/10.1093/nar/gky1003
31. Elsik CG, Tayal A, Unni DR et al (2018) Hymenoptera genome database: using HymenopteraMine to enhance genomic studies of hymenopteran insects. Methods Mol Biol 1757:513–556. https://doi.org/10.1007/978-1-4939-7737-6_17
32. Poelchau MF, Chen MJM, Lin YY, Childers CP (2018) Navigating the i5k workspace@NAL: a resource for arthropod genomes. Methods Mol Biol 1757:557–577. https://doi.org/10.1007/978-1-4939-7737-6_18
33. Sayers EW, Agarwala R, Bolton EE et al (2019) Database resources of the National Center for Biotechnology Information. Nucleic Acids Res 47:D23–D28. https://doi.org/10.1093/nar/gky1069
34. Giraldo-Calderón GI, Emrich SJ, MacCallum RM et al (2015) VectorBase: an updated bioinformatics resource for invertebrate vectors and other organisms related with human diseases. Nucleic Acids Res 43:D707–D713. https://doi.org/10.1093/nar/gku1117
35. Attardo GM, Abd-Alla AMM, Acosta-Serrano A, et al (2019) Comparative genomic analysis of six *Glossina* genomes, vectors of African trypanosomes. Genome Biol 20:187. https://doi.org/10.1186/s13059-019-1768-2
36. Li M, Christen JM, Dittmer NT et al (2018) The *Manduca sexta* serpinome: analysis of serpin genes and proteins in the tobacco hornworm. Insect Biochem Mol Biol 102:21–30. https://doi.org/10.1016/j.ibmb.2018.09.008
37. Adelman ZN, Myles KM (2018) The C-type lectin domain gene family in *Aedes aegypti* and their role in arbovirus infection. Viruses 10:367. https://doi.org/10.3390/v10070367
38. Matetovici I, Van Den Abbeele J (2018) Thioester-containing proteins in the tsetse fly (*Glossina*) and their response to trypanosome infection. Insect Mol Biol 27:414–428. https://doi.org/10.1111/imb.12382
39. Yang L, Mei Y, Fang Q et al (2017) Identification and characterization of serine protease inhibitors in a parasitic wasp, *Pteromalus puparum*. Sci Rep 7:15755. https://doi.org/10.1038/s41598-017-16000-5
40. Neafsey DE, Waterhouse RM, Abai MR et al (2015) Highly evolvable malaria vectors: the genomes of 16 *Anopheles* mosquitoes. Science 347:1258522. https://doi.org/10.1126/science.1258522
41. Terrapon N, Li C, Robertson HM et al (2014) Molecular traces of alternative social

organization in a termite genome. Nat Commun 5:3636. https://doi.org/10.1038/ncomms4636

42. Brucker RM, Funkhouser LJ, Setia S et al (2012) Insect innate immunity database (IIID): an annotation tool for identifying immune genes in insect genomes. PLoS One 7:e45125. https://doi.org/10.1371/journal.pone.0045125
43. Bryant B, Ungerer MC, Liu Q et al (2010) A caspase-like decoy molecule enhances the activity of a paralogous caspase in the yellow fever mosquito, *Aedes aegypti*. Insect Biochem Mol Biol 40:516–523. https://doi.org/10.1016/j.ibmb.2010.04.011
44. Waterhouse RM, Povelones M, Christophides GK (2010) Sequence-structure-function relations of the mosquito leucine-rich repeat immune proteins. BMC Genomics 11:531. https://doi.org/10.1186/1471-2164-11-531
45. Zou Z, Picheng Z, Weng H et al (2009) A comparative analysis of serpin genes in the silkworm genome. Genomics 93:367–375. https://doi.org/10.1016/j.ygeno.2008.12.010
46. Camacho C, Coulouris G, Avagyan V et al (2009) BLAST+: architecture and applications. BMC Bioinformatics 10:421. https://doi.org/10.1186/1471-2105-10-421
47. Jones P, Binns D, Chang HY et al (2014) InterProScan 5: genome-scale protein function classification. Bioinformatics 30:1236–1240. https://doi.org/10.1093/bioinformatics/btu031
48. Mitchell AL, Attwood TK, Babbitt PC et al (2019) InterPro in 2019: improving coverage, classification and access to protein sequence annotations. Nucleic Acids Res 47:D351–D360. https://doi.org/10.1093/nar/gky1100
49. Löytynoja A (2014) Phylogeny-aware alignment with PRANK. Methods Mol Biol 1079:155–170. https://doi.org/10.1007/978-1-62703-646-7_10
50. Katoh K, Standley DM (2013) MAFFT multiple sequence alignment software version 7: improvements in performance and usability. Mol Biol Evol 30:772–780. https://doi.org/10.1093/molbev/mst010
51. Eddy SR (2011) Accelerated profile HMM searches. PLoS Comput Biol 7:e1002195. https://doi.org/10.1371/journal.pcbi.1002195
52. Waterhouse RM (2015) A maturing understanding of the composition of the insect gene repertoire. Curr Opin Insect Sci 7:15–23. https://doi.org/10.1016/j.cois.2015.01.004
53. Waterhouse RM, Seppey M, Simão FA et al (2018) BUSCO applications from quality assessments to gene prediction and phylogenomics. Mol Biol Evol 35:543–548. https://doi.org/10.1093/molbev/msx319
54. Waterhouse RM, Seppey M, Simão FA, Zdobnov EM (2019) Using BUSCO to assess insect genomic resources. In: Methods Mol. Biol. Humana Press, New York, NY, pp 59–74
55. Pennisi E (2000) Ideas fly at gene-finding jamboree. Science 287:2182–2184. https://doi.org/10.1126/science.287.5461.2182
56. Saha S, Hosmani PS, Villalobos-Ayala K et al (2017) Improved annotation of the insect vector of citrus greening disease: biocuration by a diverse genomics community. Database (Oxford) 2017. https://doi.org/10.1093/database/bax032
57. Hosmani PS, Shippy T, Miller S et al (2019) A quick guide for student-driven community genome annotation. PLoS Comput Biol 15:e1006682. https://doi.org/10.1371/journal.pcbi.1006682
58. Ioannidis P, Simao FA, Waterhouse RM et al (2017) Genomic features of the damselfly *Calopteryx splendens* representing a sister clade to most insect orders. Genome Biol Evol 9:415–430. https://doi.org/10.1093/gbe/evx006
59. Buels R, Yao E, Diesh CM et al (2016) JBrowse: a dynamic web platform for genome visualization and analysis. Genome Biol 17:66. https://doi.org/10.1186/s13059-016-0924-1
60. Dunn NA, Unni DR, Diesh C et al (2019) Apollo: Democratizing genome annotation. PLoS Comput Biol 15:e1006790. https://doi.org/10.1371/journal.pcbi.1006790
61. Priyam A, Woodcroft BJ, Rai V, et al (2019) Sequenceserver: a modern graphical user interface for custom BLAST databases. Mol Biol Evol 36:2922–2924. https://doi.org/10.1093/molbev/msz185
62. Kim D, Langmead B, Salzberg SL (2015) HISAT: a fast spliced aligner with low memory requirements. Nat Methods 12:357–360. https://doi.org/10.1038/nmeth.3317
63. Dobin A, Davis CA, Schlesinger F et al (2013) STAR: ultrafast universal RNA-seq aligner. Bioinformatics 29:15–21. https://doi.org/10.1093/bioinformatics/bts635
64. Ramírez F, Ryan DP, Grüning B et al (2016) deepTools2: a next generation web server for deep-sequencing data analysis. Nucleic Acids Res 44:W160–W165. https://doi.org/10.1093/nar/gkw257
65. Altenhoff AM, Dessimoz C (2012) Inferring orthology and paralogy. Methods Mol Biol

855:259–279. https://doi.org/10.1007/978-1-61779-582-4_9

66. Hall BG (2013) Building phylogenetic trees from molecular data with MEGA. Mol Biol Evol 30(5):1229–1235. https://doi.org/10.1093/molbev/mst012
67. Conesa A, Madrigal P, Tarazona S et al (2016) A survey of best practices for RNA-seq data analysis. Genome Biol 17:13. https://doi.org/10.1186/s13059-016-0881-8
68. Crawford JE, Guelbeogo WM, Sanou A et al (2010) De novo transcriptome sequencing in *Anopheles funestus* using Illumina RNA-seq technology. PLoS One 5:e14202. https://doi.org/10.1371/journal.pone.0014202
69. Haas BJ, Papanicolaou A, Yassour M et al (2013) De novo transcript sequence reconstruction from RNA-seq using the trinity platform for reference generation and analysis. Nat Protoc 8:1494–1512. https://doi.org/10.1038/nprot.2013.084
70. Holt C, Yandell M (2011) MAKER2: an annotation pipeline and genome-database management tool for second-generation genome projects. BMC Bioinformatics 12:491. https://doi.org/10.1186/1471-2105-12-491
71. Ching T, Huang S, Garmire LX (2014) Power analysis and sample size estimation for RNA-Seq differential expression. RNA 20:1684–1696. https://doi.org/10.1261/rna.046011.114
72. Schurch NJ, Schofield P, Gierliński M et al (2016) How many biological replicates are needed in an RNA-seq experiment and which differential expression tool should you use? RNA 22:839–851. https://doi.org/10.1261/rna.053959.115
73. Robles JA, Qureshi SE, Stephen SJ et al (2012) Efficient experimental design and analysis strategies for the detection of differential expression using RNA-sequencing. BMC Genomics 13:484. https://doi.org/10.1186/1471-2164-13-484
74. Todd EV, Black MA, Gemmell NJ (2016) The power and promise of RNA-seq in ecology and evolution. Mol Ecol 25:1224–1241. https://doi.org/10.1111/mec.13526
75. Khalil S, Jacobson E, Chambers MC, Lazzaro BP (2015) Systemic bacterial infection and immune defense phenotypes in *Drosophila melanogaster*. J Vis Exp 99:e52613. https://doi.org/10.3791/52613
76. Taylor K, Kimbrell DA (2007) Host immune response and differential survival of the sexes in *Drosophila*. Fly (Austin) 1:197–204. https://doi.org/10.4161/fly.5082
77. Schlüns H, Sadd BM, Schmid-Hempel P, Crozier RH (2010) Infection with the trypanosome *Crithidia bombi* and expression of immune-related genes in the bumblebee *Bombus terrestris*. Dev Comp Immunol 34:705–709. https://doi.org/10.1016/j.dci.2010.02.002
78. Bray NL, Pimentel H, Melsted P, Pachter L (2016) Near-optimal probabilistic RNA-seq quantification. Nat Biotechnol 34:525–527. https://doi.org/10.1038/nbt.3519
79. Pimentel H, Bray NL, Puente S et al (2017) Differential analysis of RNA-seq incorporating quantification uncertainty. Nat Methods 14:687–690. https://doi.org/10.1038/nmeth.4324
80. Mussabekova A, Daeffler L, Imler J-L (2017) Innate and intrinsic antiviral immunity in *Drosophila*. Cell Mol Life Sci 74:2039–2054. https://doi.org/10.1007/s00018-017-2453-9
81. Myllymäki H, Valanne S, Rämet M (2014) The *Drosophila* imd signaling pathway. J Immunol 192:3455–3462. https://doi.org/10.4049/jimmunol.1303309
82. Myllymäki H, Rämet M (2014) JAK/STAT pathway in *Drosophila* immunity. Scand J Immunol 79:377–385. https://doi.org/10.1111/sji.12170
83. Valanne S, Wang J-H, Rämet M (2011) The *Drosophila* toll signaling pathway. J Immunol 186:649–656. https://doi.org/10.4049/jimmunol.1002302
84. Bosch PS, Makhijani K, Herboso L, et al (2019) Blood cells of adult *Drosophila* do not expand, but control survival after bacterial infection by induction of Drosocin around their reservoir at the respiratory epithelia. bioRxiv 578864. https://doi.org/10.1101/578864
85. Buchon N, Poidevin M, Kwon H-M et al (2009) A single modular serine protease integrates signals from pattern-recognition receptors upstream of the *Drosophila* toll pathway. Proc Natl Acad Sci 106:12442–12447. https://doi.org/10.1073/pnas.0901924106
86. Apidianakis Y, Mindrinos MN, Xiao W et al (2005) Profiling early infection responses: *Pseudomonas aeruginosa* eludes host defenses by suppressing antimicrobial peptide gene expression. Proc Natl Acad Sci 102:2573–2578. https://doi.org/10.1073/pnas.0409588102
87. Sackton TB, Werren JH, Clark AG (2013) Characterizing the infection-induced transcriptome of *Nasonia vitripennis* reveals a preponderance of taxonomically-restricted immune

genes. PLoS One 8:e83984. https://doi.org/10.1371/journal.pone.0083984

88. Gupta SK, Kupper M, Ratzka C et al (2015) Scrutinizing the immune defence inventory of *Camponotus floridanus* applying total transcriptome sequencing. BMC Genomics 16:540. https://doi.org/10.1186/s12864-015-1748-1
89. Doublet V, Poeschl Y, Gogol-Döring A et al (2017) Unity in defence: honeybee workers exhibit conserved molecular responses to diverse pathogens. BMC Genomics 18:207. https://doi.org/10.1186/s12864-017-3597-6
90. Baruzzo G, Hayer KE, Kim EJ et al (2017) Simulation-based comprehensive benchmarking of RNA-seq aligners. Nat Methods 14:135–139. https://doi.org/10.1038/nmeth.4106
91. Yi L, Pimentel H, Bray NL, Pachter L (2018) Gene-level differential analysis at transcript-level resolution. Genome Biol 19:53. https://doi.org/10.1186/s13059-018-1419-z
92. Soneson C, Love MI, Robinson MD (2016) Differential analyses for RNA-seq: transcript-level estimates improve gene-level inferences. F1000Res 4:1521. https://doi.org/10.12688/f1000research.7563.2
93. Patro R, Duggal G, Love MI et al (2017) Salmon provides fast and bias-aware quantification of transcript expression. Nat Methods 14:417–419. https://doi.org/10.1038/nmeth.4197
94. Li B, Dewey CN (2011) RSEM: accurate transcript quantification from RNA-Seq data with or without a reference genome. BMC Bioinformatics 12:323. https://doi.org/10.1186/1471-2105-12-323
95. Bolger AM, Lohse M, Usadel B (2014) Trimmomatic: a flexible trimmer for Illumina sequence data. Bioinformatics 30:2114–2120. https://doi.org/10.1093/bioinformatics/btu170
96. Gaspar JM (2018) NGmerge: merging paired-end reads via novel empirically-derived models of sequencing errors. BMC Bioinformatics 19:536. https://doi.org/10.1186/s12859-018-2579-2
97. Love MI, Huber W, Anders S (2014) Moderated estimation of fold change and dispersion for RNA-seq data with DESeq2. Genome Biol 15:550. https://doi.org/10.1186/s13059-014-0550-8
98. Law CW, Chen Y, Shi W, Smyth GK (2014) Voom: precision weights unlock linear model analysis tools for RNA-seq read counts. Genome Biol 15:R29. https://doi.org/10.1186/gb-2014-15-2-r29
99. Robinson MD, McCarthy DJ, Smyth GK (2010) edgeR: a bioconductor package for differential expression analysis of digital gene expression data. Bioinformatics 26:139–140. https://doi.org/10.1093/bioinformatics/btp616

Chapter 2

Analyzing Immunity in Non-model Insects Using De Novo Transcriptomics

Shulin He, Paul R. Johnston, and Dino P. McMahon

Abstract

With the advent of widely accessible and cost-effective next-generation sequencing technologies, it has become increasingly feasible to study insect immunity on a deep genomic or transcriptomic level. Here we introduce a protocol that is aimed at exploiting transcriptomic data to study immunity in non-model insect organisms. We provide instructions for an entire workflow, starting with successful extraction of insect RNA through to bioinformatic guidelines for the effective analysis of mRNA sequencing data. The RNA extraction procedure is based on TRIzol Reagent and a spin-column clean-up step. The bioinformatic pipeline is intended to help users identify immune genes from de novo transcriptome data and includes guidelines for conducting differential gene expression analyses on transcriptomic data. The immune gene prediction method is based on inferring protein homologs with HMMER and Blastp and takes advantage of the ImmunoDB database, which is a valuable resource for research on insect immune-related genes and gene families. The differential gene expression analysis procedure utilizes the DESeq2 package as implemented in R. We hope this protocol will serve as a useful resource for researchers aiming to study immunity in non-model insect species.

Key words RNA extraction, mRNA-seq, ImmunoDB, Immune gene prediction, Differential gene expression analysis, De novo assembly

1 Introduction

The animal immune system acts as a key interface between hosts and microbes, yet little is known about immunity in many insect species. This represents a significant gap in knowledge given that insects comprise the most species-rich group of animals by some distance [1] and that many insects such as cockroaches, beetles, and bugs can thrive in microbially diverse environments. These factors make insect immunity an attractive target for research, but relatively little is known outside of a handful of insect model species, most notably flies and beetles (e.g., [2–7]). Innate immune

Paul R. Johnston and Dino P. McMahon contributed equally to this work.

Federica Sandrelli and Gianluca Tettamanti (eds.), *Immunity in Insects*, Springer Protocols Handbooks,
https://doi.org/10.1007/978-1-0716-0259-1_2, © Springer Science+Business Media, LLC, part of Springer Nature 2020

molecules in insects occur as three broad types (not withstanding exceptions): receptors, signaling components, and effectors [3, 8]. Following infection, pattern recognition receptors bind to microorganisms, which leads to the induction of three principal signaling pathways responsible for the regulation of the insect humoral immune response, known as the Toll, immune deficiency (IMD), and Janus kinase/signal transducers and activators of transcription (JAK-STAT) pathways [9].

With the advent of next-generation sequencing, it is becoming increasingly feasible to expand our knowledge of insect immunity to non-model insect species. RNA sequencing in particular has become a popular approach in the study of immunity as it not only assists with the identification and annotation of immune genes but also can be used to explore immune function via quantitative gene expression analysis.

Several methods have been developed to annotate genes, mainly based on sequence-sequence or sequence-profile alignments to infer the homology of genes or proteins (e.g., [10–14]). The sequence–sequence alignment methods usually adapt a well-annotated genome from closely related species or public databases, such as the nr database from NCBI and Swiss-Prot from UniProt [15], in order to identify the most similar sequences. The sequence-profile methods adapt aligned gene/protein families, such as pfam [16], to infer homologous sequences. These methods are based on a position-specific score matrix and are more sensitive than methods based on sequence–sequence similarity [11, 17]. As some immune-related gene families share important domains in their proteins, we recommend using both strategies to identify immune-related genes. A few databases on insect immune gene families have been constructed such as ImmunoDB [18] and Insect Innate Immunity Database [19].

Methods for the quantification of gene expression are based on alignment of raw reads to either a genome, accounting for overlapping genomic features (specified by an accompanying annotation) (e.g., [20, 21]), or a transcriptome (from a well-annotated genome or a de novo RNA-seq assembly) and then summing counts at the gene level (e.g., [22–24]). We recommend the latter due to the superior control of false discovery rates for genes that show only differential usage of isoforms of different length [25]. Modern transcript-level quantification programs adapt efficient pseudo-alignment approaches [22, 24] which show similar accuracy to more computationally demanding alignment-based methods [22, 23]. This allows the analysis to be performed on a standard personal computer in a short time [22]. Crucially, accurate transcript-level quantification can be applied in both model and non-model organisms and can be carried out without a reference genome.

In differential expressional analysis, expression data are usually modeled by using the negative binomial distribution [26, 27] or, in some cases, by using linear modeling of transformed counts with observation-level weights [28]. We recommend using the DESeq2 package [26] or edgeR [27] in the R language environment [29]. Both are appropriate for analyzing estimated gene-level counts derived from transcript-level quantification and produce similar results. The R package "tximport" is usually used to facilitate the reading and aggregation of count data for modeling [25].

In this chapter, we describe RNA extraction and preparation for next-generation sequencing and analysis of sequencing data. The RNA extraction method takes advantage of TRIzol Reagent and clean-up steps using an RNeasy Mini Kit. We do not include experimental guidelines and assume that users plan to sample insect tissues that are appropriate for the particular research question being asked, although clear patterns of differential gene expression rely on a well-designed experimental setup with a sufficient degree of replication. We also do not include installation instructions for the programs, and we encourage users to consult the instruction files in the programs. The users should have a working knowledge of Linux and command-line interface.

2 Materials

2.1 Equipment and Materials

1. 50–100 mg insect tissues (*see* **Note 1**).
2. 80 °C freezer.
3. Tissue homogenizer (e.g., MP Bio FastPrep 24 Homogenizer).
4. Fume hood.
5. Microcentrifuge (requires 13,000 × *g*, cold to 4 °C).
6. Fluorometer (e.g., Invitrogen Qubit 4 Fluorometer).
7. Agilent Bioanalyzer 2100.
8. Microvolume spectrophotometer (e.g., NanoDrop).
9. Heat block (up to 70 °C).
10. Vortexer.
11. Forceps.
12. Liquid N Dewar flask.
13. Ice box.
14. 2-ml tube racks.
15. Stainless steel beads (5 mm).
16. Pipettes (0.5–10 μl, 2–20 μl, 20–1000 μl).
17. Sterile, RNase-/DNase-free pipette tips with filters (0.1–20 μl, 2–200 μl, 100–1000 μl).
18. Sterile, RNase-free disposable microcentrifuge tubes (1.5 ml).

2.2 Reagent Kits

1. RNeasy® Mini Kit (Qiagen).
2. TURBO™ DNase (Invitrogen).
3. THE RNA Storage Solution (Invitrogen).
4. Qubit dsDNA HS Assay Kit and Qubit RNA HS Assay Kit (Invitrogen).
5. RNA 6000 Nano Kit (Agilent Genomics).

2.3 Chemicals

1. Ice.
2. Liquid nitrogen.
3. Absolute ethanol (molecular biology grade).
4. RNase-free water.
5. TRIzol Reagent.
6. Chloroform (molecular biology grade).
7. Isopropanol (molecular biology grade).
8. 75% ethanol (molecular biology grade).
9. RNase AWAY® (or similar) surface decontaminant.
10. Diethyl pyrocarbonate (DEPC) (*see* **Note 2**).

2.4 Solutions

1. DEPC water: adding DEPC to Milli-Q water to give a 0.1% v/v solution, close the bottle, shake carefully to mix thoroughly, and incubate overnight at room temperature. Autoclave treated water to remove DEPC traces at least 15 min (*see* **Note 3**).

3 Methods

3.1 RNA Extraction

3.1.1 Sample Preparation

1. The insect tissues are collected in a 1.5-ml tube, flash frozen in liquid nitrogen, and stored immediately at −80 °C.

3.1.2 RNA Extraction

The method presented below was successfully used in our laboratory to extract RNA from whole bodies of cockroaches and termites. We homogenized insect tissues with a homogenizer, extracted total RNA with TRIzol Reagent, and removed DNA with DNase followed by a clean-up using the RNeasy Mini Kit (Qiagen). Samples were eluted in RNase-free water or elution buffer.

2. Workspace preparation for RNA extraction (*see* **Note 4**).
3. Add a stainless-steel bead and 1 ml of TRIzol Reagent per 50–100 mg of tissue to the sample in a fume cupboard; be careful when handling TRIzol (*see* **Note 5**).
4. Homogenize samples using a homogenizer (*see* **Note 6**).
5. Incubate for 5 min at room temperature to allow complete dissociation of the nucleoproteins complex.

6. Add 0.2 ml of chloroform per 1 ml TRIzol, cap the tube carefully, and shake vigorously for 15 s (*see* **Note 7**).
7. Incubate for 2–3 min at room temperature.
8. Centrifuge for 15 min at 12,000 × *g* at 4 °C.
9. Transfer the upper aqueous phase containing the RNA to a new tube (*see* **Note 8**).
10. Add 0.5 ml of isopropanol to the transferred aqueous phase per 1 ml TRIzol. Gently mix the solution (*see* **Note 9**).
11. Incubate for 10 min at room temperature.
12. Centrifuge for 10 min at 12,000 × *g* at 4 °C.
13. Discard the supernatant with a micropipette.
14. Resuspend the pellet in 1 ml of 75% ethanol per 1 ml TRIzol.
15. Vortex the sample briefly, and then centrifuge for 5 min at 7500 × *g* at 4 °C.
16. Discard the supernatant with a micropipette.
17. Air-dry the RNA pellet for 5–10 min.
18. Resuspend the pellet in 89 μl of THE RNA Storage Solution. Ensure the RNA resolves in solution thoroughly by gentle pipetting (*see* **Note 10**).
19. Add 10 μl 10 × TURBO DNase buffer and 1 μl TURBO DNase to RNA solution.
20. Incubate at 37 °C for 30 min.
21. Add 350 μl Buffer RLT(RNeasy® Mini Kit, Qiagen), and mix well.
22. Add 250 μl ethanol (96–100%) to the diluted RNA and mix well by pipetting.
23. Transfer the sample (700 μl) to an RNeasy Mini spin column placed in a 2-ml collection tube (RNeasy® Mini Kit, Qiagen). Close the lid, and centrifuge for 15 s at 8000 × *g* at room temperature. Discard the flow-through.
24. Add 500 μl Buffer RPE (RNeasy® Mini Kit, Qiagen) to the spin column. Close the lid, and centrifuge for 15 s at 8000 × *g* at room temperature. Discard the flow-through.
25. Add 500 μl Buffer RPE (RNeasy® Mini Kit, Qiagen) to the RNeasy spin column. Close the lid gently, and centrifuge for 2 min at 8000 × *g* at room temperature.
26. Place the RNeasy spin column in a new 2-ml collection tube (RNeasy® Mini Kit, Qiagen), and centrifuge at full speed for 1 min at room temperature.
27. Place the RNeasy spin column in a new 1.5-ml collection tube (RNeasy® Mini Kit, Qiagen). Add 30 μl RNase-free water directly to the spin column membrane. Close the lid,

and centrifuge for 1 min at 8000 × *g* at room temperature (*see* **Note 11**).

28. Repeat elution step using the eluate and the collection tube from **step 27**.
29. Aliquot purified RNA into three volumes before storing at −80 °C: 4 μl for quantification and basic quality checks, 2 μl for quality checking using the Bioanalyzer, and the rest for sequencing.

3.1.3 RNA Quality Check

The following steps are used to determine the concentration, quality, and integrity of RNA (*see* **Note 12**):

1. Defrost the aliquot for concentration and quality checking on ice.
2. Measure the RNA quality on NanoDrop system (*see* **Note 13**).
3. Measure the RNA and DNA concentrations using Qubit RNA HS Assay Kit and Qubit dsDNA HS Assay Kit with the Invitrogen Qubit 4 Fluorometer.
4. Prepare the reagents in the Agilent RNA 6000 Nano Kit, following manufacturer's instructions.
5. Defrost those aliquots to be checked on ice.
6. Measure the quality following manufacturer's instructions (*see* Fig. 1, **Note 14**).
7. Send frozen total RNA on dry ice to the sequencing facility if it has passed the quality check step (*see* **Note 15**).

3.2 RNA Sequencing

Library preparation often includes the steps of mRNA isolation, fragmentation, first strand synthesis, and second strand synthesis. A number of commercial kits provide all reagents and instructions to perform these steps (e.g., Bioo Scientific, TruSeq, Illumina). Specific steps may vary depending upon the sequencing instrument used. While many methods are available to prepare RNA-seq libraries, the quality and accurate quantitation of the total RNA extract remain paramount to success.

3.3 RNA Sequence Data Analysis

The following bioinformatic analysis is suitable for target insects or their closely related species that do not have an available genome. If a reference transcriptome from a high-quality reference genome is available, we recommend following the protocols described in Chapter 1 of this book ("Characterization of Insect Immune Systems from Genomic Data" by Waterhouse and colleagues), which provides details for the analysis of insect immune systems from genomic data.

The pipeline described here contains instructions for de novo assembly of a reference transcriptome, immune-related genes prediction, and differential gene expression. Firstly, we build up a

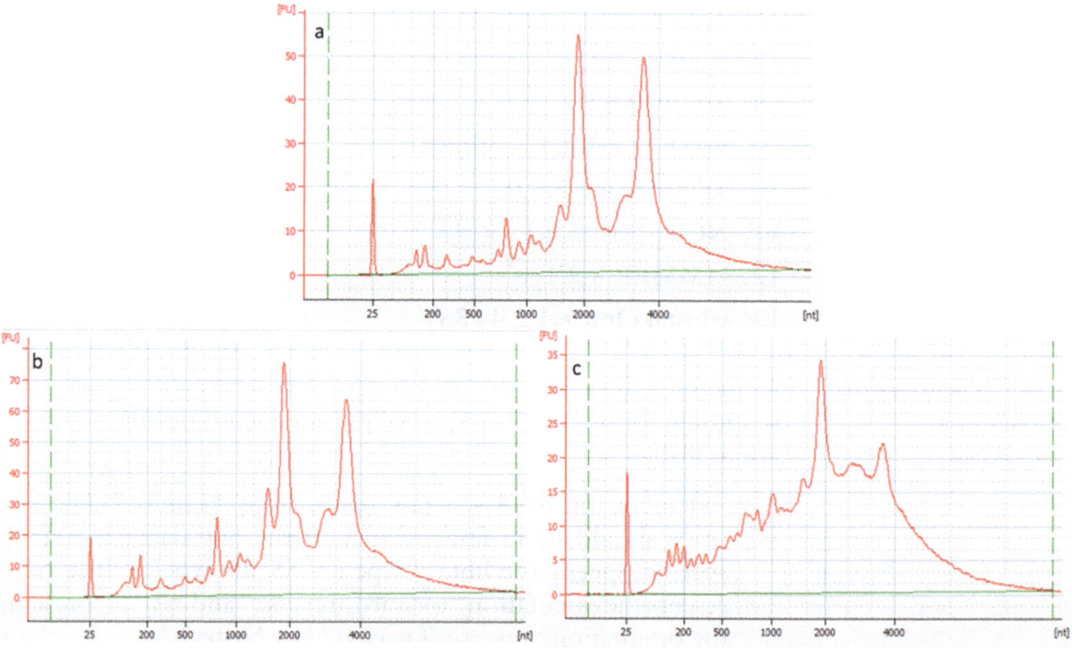

Fig. 1 Bioanalyzer example traces of RNA extracts from termite soldiers and cockroaches. (**a** and **b**) Good traces (termite soldiers), following methods as described in this protocol, resulting in RIN values of 7.50 and 6.80, respectively. (**c**) Poor trace (cockroach), resulting from thawing stored RNA once prior to extraction, leading to RNA degradation, with an RIN of 4.00

reference transcriptome by using Trinity [30, 31], which is a robust assembler for constructing de novo transcriptomes with de Bruijn graphs and provides cluster information in the output. Secondly, we provide guidelines to identify immune-related proteins using TransDecoder [32], which uses the predicted proteins from the Trinity output and takes advantages of well-annotated ImmunoDB [18] and UniProt [15] databases. The identification steps employ strategies based on sequence-profile alignments as implemented in HMMER (http://hmmer.org/) and sequence-sequence alignments via Blastp [10]. Following identification, a manual curation step is required to minimize false positives resulting from transcript fragmentation in transcriptomic data. As immune gene identification relies on annotated immune proteins, this protocol is not suitable for the discovery of novel immune genes. Finally, we provide instructions for the analysis of differential gene expression by using Salmon [24] and DESeq2 [26] in R [29].

3.3.1 Software (See Note 16)

1. R 3.4.3.
2. Salmon 0.13.1.
3. Readr 1.1.0.
4. Tximport 1.4.0.
5. DESeq2 1.16.1.

6. Tidyr 0.6.1.
7. ggplot2 2.2.1.
8. Trinity 2.6.5.
9. Bowtie2 2.3.4.1 (optional).
10. Transdecoder5.5.0.
11. NCBI-blast 2.8.1.
12. HMMER 3.2.1.
13. Clustal Omega1.2.4 [33].

3.3.2 De Novo Assembly of a Reference Transcriptome

As mentioned before, this step is necessary when a reference transcriptome for insect species, usually from a genome annotation, is not available.

1. Assemble the RNA-seq raw reads using "Trinity." The raw reads should be in fastq format. If the raw data are strand-specific, specify the library type ("--SS_lib_type"): "FR" represents the orientation "forward/reverse" and "RF" represents the orientation "reverse/forward" (*see* **Note 17**).

```
Trinity --seqTypefq --max_memory 32G \
--left    lib_1_S1_R1_001.fastq.gz,lib_2_S2_R1_001.
fastq.gz \
--right   lib_1_S1_R2_001.fastq.gz,lib_2_S2_R2_001.
fastq.gz \
--CPU  12  --trimmomatic  --SS_lib_type  FR  --full_
cleanup \
--output trinity_assembly&> assembly_log01.txt
```

   #Copy this file to a new directory before abundances quantifying:

```
mkdir quant  &&cd quant  &&cp ../trinity_assembly/
Trinity.fasta transcriptome.fa
```

2. Generate a file relating transcripts to genes.

   #Generate a file that maps genes to transcripts:

```
perl $TRINITY_HOME/util/support_scripts/get_Trinity_
gene_to_trans_map.pl \
transcriptome.fa> gene2tx.tsv
```

   #Create a csv file with headers (transcripts to genes):

```
awk 'BEGIN{print "TXNAME,GENEID" }'> tx2gene.csv
#Write transcripts/genes data to the csv file:
awk -F $'\t'' {t = $1; $1 = $2; $2 = t; print;} '
OFS=$',' gene2tx.tsv >> tx2gene.csv
```

3.3.3 Immune Protein Identification

This step is to predict immune proteins by applying hmmsearch and blastp. ImmunoDB is a useful resource that provides information on immune genes and gene families across different insect species and even a few other arthropod species [18].

1. Predict proteins from assembly using TransDecoder (*see* **Note 18**).

```
TransDecoder.LongOrfs -t transcriptome.fa -m
60 &&TransDecoder.Predict \.
-t transcriptome.fa
```

2. Download immune gene family sequences from ImmunoDB database at the insect level, store sequences for each family in a FASTA file, and name the file with the family name (*see* **Note 19**).
3. Build HMM search profiles and the Blast database.

(a) HMM profile and Blast database building.

#Align all sequences for each gene family:

```
clustalo -iEACH_IMMUNE_FAMILY.fa \
-oEACH_IMMUNE_FAMILY.clustalo.alin --outfmt=st
```

#Edit the header of alignment files and merge them into a single file:

```
sed "1a\#=GF ID IMMUNE_FAMILY_NAME" EACH_IMMUNE_
FAMILY.clustalo.alin \
  >>immunedb.sto
```

#Build HMM profiles based on above alignments:

```
hmmbuild immunedb.hmm immunedb.sto
```

#Combine sequences of all immune gene families and build a Blast database:

```
cat *.fasta>immunedb.fasta&&makeblastdb -in immunedb.
fa -dbtype prot
```

#As sequence names from ImmunoDB are not suitable for further analysis, we generated a file containing sequence IDs and their corresponding gene family names:

```
awk '$1~/^>/{gsub(">","",$1); print $1"\t""
IMMUNE_FAMILY_NAME "}' \
EACH_IMMUNE_FAMILY.fa>> seq2family
```

(b) UniProtKB/Swiss-Prot database preparation.

#Download reviewed UniProtKB/Swiss-Prot database:

```
wget ftp://ftp.uniprot.org/pub/databases/
uniprot/current_release/knowledgebase/com-
plete/uniprot_sprot.fasta.gz
```

#Unzip sequences and make a Blast database:

```
gunzip uniprot_sprot.fasta.gz &&makeblastdb -in uni-
prot_sprot.fasta -dbtype prot
```

4. Predict immune proteins based on HMM profiles and the immune Blast database. As a conservative measure, we only consider proteins for subsequent curation when the same immune gene family is predicted from both approaches.

```
hmmsearch --tblouttarget_out.tab --noali --notextw -E 1e-5 --domE 1e-3 \
--incE 0.001 --cpu 4 immunedb.hmm transcript.fa.pep
blastp -query transcript.fa.pep -dbimmunedb.fa -num_threads 4 -evalue 1e-5 \
-max_target_seqs 1 -outfmt 6 -out transcript_immunedb.blastp.outfmt6
#Edit the hmmsearch output to increase readability:
sed '/^#/d' target_out.tab |sed 's/ */ /g'|cut -f1,3,5,8 -d " " |sed 's/ ^t/g' >hmm.out
#Change the blastp target names and choose the immune proteins that are identified by both search strategies to belong to the same immune family:
awk 'NR==FNR{a[$1]=$2;next}{print $1 "\t" a[$2]}' seq2family \
transcript_immunedb.blastp.outfmt6 |sort -u| \
awk 'NR==FNR{a[$1]=$2;next}$2==a[$1]{print $0"\t" a[$1]}' - hmm.out \
>hmm.blastp.out
```

5. It is necessary to make a further curation with a manually reviewed database. Here, we suggest to blast TransDecoder predicted proteins against the UniProtKB/Swiss-Prot database.

```
blastp -query protein.fa -dbuniprot_sprot.fasta
-num_threads 4 -evalue 1e-5 \
-max_target_seqs 1 \
-outfmt "6 qseqid sseqid qlen qstart qend slens start
send bitscore evalue stitle" \
-out assembly.uniprot.blastp.outfmt6
awk-F $'\t' 'NR==FNR{a[$1]=$NF;next}{print $0"\t"
a[$1]}' assembly.uniprot.blastp.outfmt6 \
hmm.out>hmm.curation
```

   Confirm the annotation of immune proteins by comparing output from two different databases manually. This step requires some immunity-related background as the protein names from UniProtKB/Swiss-Prot may be different from the immune gene family name (*see* **Note 20**).

6. Retrieve the immune gene IDs based on the transcripts to genes mapping file (*see* **Note 21**).

3.3.4 Quantify Transcript Abundances

1. Index the transcripts using "salmon index" (*see* **Note 22**).

```
salmon index -t transcriptome.fa -itx_index
```

2. Quantify abundances for each sample using "salmon quant" (*see* **Note 23**).

```
salmon quant -itx_index -l A \
-1 treated_rep1_1.fastq.gz -2 treated_rep1_2.fastq.gz \
-p 4 -o treated_rep1
```

3.3.5 Differential Expression Analysis in R

1. Begin a new R session and load the required libraries. The supplemental code assumes that the working directory of the R session contains the salmon output.

```
library(readr)
library(tximport)
library(DESeq2)
library(tidyr)
library(ggplot2)
```

2. Import count data from the salmon output directories using R package "tximport." The names provided by "names(files)" will be used to denote the sample names and are inherited by the columns of the count matrix.

   #Construct file paths to salmon outputs:

```
files<- file.path("./", list.dirs(recursive = FALSE),
"quant.sf")
```

   #Name the files according to their salmon output directory name:

```
names(files) <- gsub("./", "", list.dirs(recursive
= FALSE))
#Check the paths to files:
all(file.exists(files))
[1] TRUE
```

   #Read the tx2gene.csv (generated at **step 2** of Subheading 3.3.2) in a data.frame linking transcript ID to gene ID:

```
tx2gene <- read.csv("tx2gene.csv")
```

   #Import the salmon output files with tximport:

```
txi<- tximport(files, type = "salmon", tx2gene =
tx2gene)
```

3. Construct a data frame to relate sample names with factors in your experimental design.

   #Here we construct a data frame that links samples to the level of "condition" factor. If samples are not named according to factors and replicates, a data frame which contains a column of sample names and columns of corresponding factors is required.

```
sampleTable<- data.frame(sample =
colnames(txi$counts))
sampleTable<- separate(sampleTable, sample, into=
c("condition", "replicate"))
wnames(sampleTable) <- colnames(txi$counts)
```

4. Create a DESeqDataSet object containing the transcript counts, sample information data frame, and design formula.

```
dds<- DESeqDataSetFromTximport(txi, sampleTable,
~condition)
```

5. Run DESeq to estimate size factors and dispersions (*see* **Note 24**).

```
dds<- DESeq(dds)
```

6. Extract desired results from DESeq analysis. If there are more than two levels in a factor, extract the results of a specific contrast of two levels of a factor by using the "contrast" argument in the "results" function (*see* **Note 25**).

```
res_treat_control<-results(dds, contrast =
c("condition", "treated", "control"), lfcThreshold
= 1, alpha = 0.05)
```

7. Visualization by using principal component analysis after regularized log transformation to remove the mean-variance dependence.

```
rld<- rlog(dds)
plotPCA(rld)
pcaData<- plotPCA(rld, intgroup = "condition", re-
turnData = TRUE)
percentVar<- round(100 * attr(pcaData, "percent-
Var"))
ggplot(pcaData, aes(PC1, PC2, color = condi-
tion)) + geom_point(size = 5, alpha = 0.75) +
xlab(paste0("PC1: ", percentVar[1], "% variance"))
+ ylab(paste0("PC2: ", percentVar[2], "% vari-
ance")) + coord_fixed()
```

8. Visualization of differentially expressed immune genes with heatmaps.

   #Read manually curated immune protein table from Subheading 3.3.3:

```
immune<-read.csv("hmm.curation",sep="\t",
header=F,stringsAsFactors=F) %>% separate
(V1, into=c("gene", "iso"),sep="_i",
remove=F)
```

   #Extract regulated immune gene IDs:

```
DEGs<-      dplyr::filter(as.data.frame(res_treat_con-
trol), padj<0.05 & |log2FoldChange| >1)
immune_regulated<-DEGs[DEGs$row %in% immune$gene,]
```

   #Produce a heatmap for differentially regulated immune genes:

```
library(pheatmap)
immune_regulated_n<-assay(rld)[as.
character(immune_regulated[,c("row")])]
pheatmap(immune_regulated_n, annotation_
col=sampleTable)
```

4 Notes

1. Clean and pure samples are essential prerequisites for acquiring reliable total RNA. After collection, freeze samples quickly using liquid nitrogen. Never thaw the samples until they are

immersed in TRIzol for extraction. Ensure that liquid nitrogen is completely evaporated before immersing samples in TRIzol Reagent to avoid nitrogen gas build-up inside tubes (*see* **Note 5**).

2. DEPC is optional. DEPC is highly toxic and must be used in consultation with product safety instructions (e.g., work under a fume hood).
3. Make DEPC in a clean glass bottle: it is recommended to autoclave the treated water twice.
4. An RNase-free workspace and general aseptic lab practice is essential for successful RNA extraction. RNase AWAY® (or similar products) can be used to clean the workspace to inactivate RNases. Wipe reusable equipment (pipettes, racks) with RNase AWAY® and rinse with DEPC water if possible before use. Ensure that the rotor of the microcentrifuge is clean. Use filter tips during extraction. Prepare ahead of schedule to allow a seamless workflow, e.g., aliquot chemicals/solutions into pre-labelled tubes where possible.
5. Be cautious when handling toxic TRIzol Reagent. Wear eye protection and lab clothing and carry out work under a fume hood at all times. Use storage and microcentrifuge tubes with screw caps or safety locks when handling TRIzol or chloroform, particularly following freezing with liquid nitrogen. Where possible, avoid storing samples immersed in TRIzol in the freezer. If unavoidable, take particular care when defrosting material as nitrogen carried over from the first freezing step may build up and explode during thawing.
6. Short-time (15–20 s) and multiple disruptions are recommended for uniform homogenization. Sufficient homogenization and lysis are critical for ensuring a good quality of RNA.
7. Handle chloroform under the fume cupboard. Make sure the tube closes tightly, and use screw lid or microcentrifuge tubes with safety locks during extractions to prevent escape of chloroform or other reagents. Thorough mixing is a crucial step to make sure the proteins are denatured.
8. Be sure to only keep the tips in the aqueous phase. It is better to collect a smaller volume than to risk contamination.
9. Once extracted, carefully mix the solution by pipetting in the tube rather than by vortexing or vigorous shaking, as this could shear the RNA.
10. It is also recommended to heat the tube in incubator at 55–60 °C for 10 min. However, this could make insect RNA look degraded during Agilent quality check.
11. Leave tube at room temperature for 2 min after adding RNase-free water.

12. It is recommended to check DNA concentration and incorporate a second DNase digest step if necessary.
13. Pure RNA should have an absorbance ratio at of ~2.0 at OD_{260nm}/OD_{280nm}. If the ratio varies much from this, protein, phenol, or other contaminants may be present.
14. The final trace gives an indication of the integrity of the total RNA. The bands for decent quality of total RNA should appear as Fig. 1a, b, with two clear peaks. The RNA integrity (RIN) value is the ratio of the large (28S) to the small (18S) ribosomal RNA subunit. It indicates the quality of total RNA as the large subunit is known to degrade quickly; intact 28S should be twice as intense as the 18S. The perfect RIN value is 10 and the higher RIN value indicates the RNA being more intact. A value of 7 indicates little degradation of the RNA sample. We found samples with lower RIN values (e.g., 6) were also suitable for sequencing. To denature the insect RNA would change the trace because insect 28s RNA contains a hidden break and it produces two similar fragments like 18s after denaturation [34]. Please note the RIN may not accurately represent the integrity of mRNA. Assessing the trace by eye may give the best indication of quality—if degradation has occurred, the trace will look closer to Fig. 1c.
15. Dehydrated total RNA can also be sent successfully compared to frozen liquid samples sent on dry ice [35].
16. The programs used here are all freely available and are updated regularly. We currently use the listed programs and versions, and these versions (or higher) are required prior to carrying out the described steps.
17. Provide appropriate arguments to "--max_memory" and "--CPU" to control resource usage and save computing time.
18. Be aware of the predicted protein length, as some interesting peptides are short (recommend 60 aa). Homolog searches during prediction can also be performed, which requires some additional homology search steps suggested in TransDecoder manual (https://github.com/TransDecoder/TransDecoder/wiki). We also recommend performing the prediction step in a new directory.
19. Be cautious when downloading gene members in immune pathways from ImmunoDB, which are not clearly separated by families but by orthologous groups. It is also reasonable to download sequences of interesting immune protein families from other resources, such as the UniProt database.
20. As the ImmunoDB database is smaller than UniProt, some immune proteins/genes, especially immune pathway members,

can end up being assigned to a different gene family name. Misidentification also could be a result of gene fragments in transcriptomic data.

21. Here we use the transcripts to genes mapping file derived from the head information of the Trinity assembly. It is also possible to cluster the transcripts in other software and retrieve the gene ID based on that.

22. The input file should be in FASTA format containing one sequence per transcript. We recommend performing this step in a new directory.

23. Quantification is performed one time per sample. Here we named the output according to the combination of the level of the factor "condition" ("treated") and the replicate number ("1") (treated_rep1—the first replicate of the treated group). We recommend naming the output directories like this according to the factors associated with the samples. Provide an appropriate parameter for the "-p" argument to match available computer resources.

24. The DESeq is a wrapped function and will (1) estimate size factors to bring counts in each sample to a common scale, (2) estimate dispersions for negative binomial GLM fitting, and (3) perform negative binomial GLM fitting and Wald statistics.

25. Differentially expressed genes are tested on the basis of p value and fold change. For example, the thresholds of FDR-corrected $p < 0.05$ and log2 fold change >1 are commonly adopted. Therefore, we include the fold change threshold in the statistical test using lfcThreshold = 1.

    An alternative is to test if the log2 fold changes equal to 0 (the default test in results function, lfcThreshold = 0) and subsequently perform a post-hoc filtering based on FDR-adjusted *p* value and log2 fold change. However, this filtering makes the *p* values difficult to interpret because they are calculated from a log2 fold change test on different threshold.

    Sometimes it is necessary to test all levels of a factor at once, e.g., all genes which vary in different levels of "condition."

    This can be performed by using the likelihood ratio test (LRT) between a full model (here ~condition) and a reduced model (here ~1).

```
dds<- DESeqDataSetFromTximport(txi, sampleTable, ~condition)

dds<- DESeq(dds, parallel = TRUE, test = "LRT", reduced = ~1)

res_1<- results(dds, alpha = 0.05)
```

References

1. Stork NE (2018) How many species of insects and other terrestrial arthropods are there on earth? Annu Rev Entomol 63:31–45
2. Buchon N, Silverman N, Cherry S (2014) Immunity in Drosophila melanogaster—from microbial recognition to whole-organism physiology. Nat Rev Immunol 14(12):796
3. Hillyer JF (2016) Insect immunology and hematopoiesis. Dev Comp Immunol 58:102–118
4. Hoffmann JA, Reichhart JM (2002) Drosophila innate immunity: an evolutionary perspective. Nat Immunol 3(2):121
5. Hoffmann JA (2003) The immune response of drosophila. Nature 426(6962):33
6. Milutinović B, Peuß R, Ferro K et al (2016) Immune priming in arthropods: an update focusing on the red flour beetle. Zoology 119(4):254–261
7. Rolff J, Reynolds S (eds) (2009) Insect infection and immunity: evolution, ecology, and mechanisms. England, Oxford
8. Viljakainen L (2015) Evolutionary genetics of insect innate immunity. Brief Funct Genomics 14(6):407–412
9. Lemaitre B, Hoffmann J (2007) The host defense of Drosophila melanogaster. Annu Rev Immunol 25:697–743
10. Altschul SF, Gish W, Miller W et al (1990) Basic local alignment search tool. J Mol Biol 215(3):403–410
11. Altschul SF, Madden TL, Schäffer AA et al (1997) Gapped BLAST and PSI-BLAST: a new generation of protein database search programs. Nucleic Acids Res 25(17):3389–3402
12. Mistry J, Finn RD, Eddy SR et al (2013) Challenges in homology search: HMMER3 and convergent evolution of coiled-coil regions. Nucleic Acids Res 41(12):e121–e121
13. Krogh A, Brown M, Mian IS et al (1994) Hidden Markov models in computational biology: applications to protein modeling. J Mol Biol 235(5):1501–1531
14. Wheeler TJ, Eddy SR (2013) Nhmmer: DNA homology search with profile HMMs. Bioinformatics 29(19):2487–2489
15. UniProt Consortium (2018) UniProt: a worldwide hub of protein knowledge. Nucleic Acids Res 47(D1):D506–D515
16. El-Gebali S, Mistry J, Bateman A et al (2018) The Pfam protein families database in 2019. Nucleic Acids Res 47(D1):D427–D432
17. Eddy SR (1998) Profile hidden Markov models. Bioinformatics 14(9):755–763
18. Waterhouse RM, Kriventseva EV, Meister S et al (2007) Evolutionary dynamics of immune-related genes and pathways in disease-vector mosquitoes. Science 316(5832):1738–1743
19. Brucker RM, Funkhouser LJ, Setia S et al (2012) Insect innate immunity database (IIID): an annotation tool for identifying immune genes in insect genomes. PLoS One 7(9):e45125
20. Anders S, Pyl PT, Huber W (2015) HTSeq-a python framework to work with high-throughput sequencing data. Bioinformatics 31(2):166–169
21. Liao Y, Smyth GK, Shi W (2013) featureCounts: an efficient general purpose program for assigning sequence reads to genomic features. Bioinformatics 30(7):923–930
22. Bray NL, Pimentel H, Melsted P et al (2016) Near-optimal probabilistic RNA-seq quantification. Nat Biotechnol 34(5):525
23. Li B, Dewey CN (2011) RSEM: accurate transcript quantification from RNA-Seq data with or without a reference genome. BMC Bioinformatics 12(1):323
24. Patro R, Duggal G, Love MI et al (2017) Salmon provides fast and bias-aware quantification of transcript expression. Nat Methods 14(4):417
25. Soneson C, Love MI, Robinson MD (2015) Differential analyses for RNA-seq: transcript-level estimates improve gene-level inferences. F1000Res 4:1521
26. Love MI, Huber W, Anders S (2014) Moderated estimation of fold change and dispersion for RNA-seq data with DESeq2. Genome Biol 15(12):550
27. Robinson MD, McCarthy DJ, Smyth GK (2010) edgeR: a bioconductor package for differential expression analysis of digital gene expression data. Bioinformatics 26(1):139–140
28. Law CW, Chen Y, Shi W et al (2014) Voom: precision weights unlock linear model analysis tools for RNA-seq read counts. Genome Biol 15(2):R29
29. Team RC (2017) R: a language and environment for statistical computing (version 3.4. 2)[computer software]. R Foundation for Statistical Computing, Vienna, Austria
30. Grabherr MG, Haas BJ, Yassour M et al (2011) Full-length transcriptome assembly from RNA-seq data without a reference genome. Nat Biotechnol 29(7):644–652
31. Haas BJ, Papanicolaou A, Yassour M et al (2013) De novo transcript sequence reconstruction from RNA-seq using the trinity

platform for reference generation and analysis. Nat Protoc 8(8):1494–1512

32. Haas BJ Papanicolaou A manuscript in prep. http://transdecoder.github.io
33. Madeira F, Lee J, Buso N et al (2019) The EMBL-EBI search and sequence analysis tools APIs in 2019. Nucleic Acids Res 47(W1):W63–W641
34. Winnebeck EC, Millar CD, Warman GR (2010) Why does insect RNA look degraded? J Insect Sci 10(1):159
35. Johnson MT, Carpenter EJ, Tian Z et al (2012) Evaluating methods for isolating total RNA and predicting the success of sequencing phylogenetically diverse plant transcriptomes. PLoS One 7(11):e50226

Chapter 3

Preparation of Insect Protein Samples for Label-Free Proteomic Quantification by LC-Mass Spectrometry

Alexandro Rodríguez-Rojas and Jens Rolff

Abstract

Insects are important biological models for the study of immune function and development. The development of proteomics and protein identification techniques combined with next-generation sequencing (NGS) for genome and transcriptome sequencing provides a powerful tool for the study of insect physiology, including insect immunity, stress biology, reproduction, the influence of environmental factors, and many other aspects of insect life. Proteomic studies are also useful to study post-translational modifications that play a fundamental role in animal physiology since a large fraction of the proteome is modified via oxidation, glycosylation, and phosphorylation. The use of proteomics to study insects offers an opportunity to advance in new directions to further our understanding of protein function and their networks. This is particularly true for the characterization of the insect immune system. The aim of this work is to provide a comprehensive methodology to prepare insect samples for proteomic analysis.

Key words Insect immunity, Proteomic, Label-free quantification, Mass spectrometry, StageTips

1 Introduction

Insects are always under attack by microbial pathogens and multicellular parasites. Their immune system provides efficient protection contributing to the evolutionary success of insects as the most speciose animal taxon [1]. Although the insect immune system is devoid of lymphocytes or antibodies, they can effectively combat infections. They possess cellular responses that include encapsulation of pathogens and phagocytosis [2]. The humoral response in insects mostly relies on the secretion of antimicrobial peptides [3], lysozymes, and protein-mediated effectors such as phenoloxidases and other chemical defenses. The recognition system entails some specificity against different types of pathogens involving receptors such as peptidoglycan recognition protein or β-glucan recognition protein that activate signaling pathways including the Toll, IMD, and JAK-STAT pathways [1].

Federica Sandrelli and Gianluca Tettamanti (eds.), *Immunity in Insects*, Springer Protocols Handbooks,
https://doi.org/10.1007/978-1-0716-0259-1_3,

All insect defense systems directly or indirectly involve proteins that change their level of expression or change from an inactive to an active form. Because all of these changes imply different protein levels or protein modification such as via phosphorylation, glycosylation, and oxidation, they are ideal to be studied using proteomic approaches. Amino acid sequence database is an essential component in the current proteomics with mass spectrometry. The development of next-generation sequencing is providing whole genome sequences or transcriptomic sequencing (RNAseq) for more insect species over time, making proteomic approaches more reachable.

Proteomic technologies have improved in recent years and the resolution and detectability now offer very good opportunities to study changes in the global proteome with a high coverage [4]. Because of the improvements of the technology, label-free quantification (i.e., no need for radioactive isotopes) has become a cost-effective option for proteomic studies [5]. Insects have an open circulatory system and their immune system proteome can be studied using the whole body, hemolymph, or specific organs such as the fatbody. In this chapter, we describe how to process samples to study the proteome of insect immune systems by processing whole insect bodies or specific tissues.

2 General Considerations on Sampling Protein from Insects

We recommend that sample should be collected as freshly as possible. If the whole body is going to be processed, the method of choice is to freeze the specimens in liquid nitrogen as we have described elsewhere [6, 7]. This freezing procedure can be applied to other tissue or insect organs in a similar way. Individual insects or groups from the same treatment can be processed together. However, if the whole body is used, even small insects can provide enough material when they are processed individually.

In the case of processing hemolymph samples, a quick extraction method to avoid activation of phenoloxidases (POs) and clotting cascades is crucial. In some cases, the use of anti-melanization agents such as ascorbic acid or phenylthiourea is unavoidable but in many cases not necessary. If melanization or clotting occurs, the samples could be unsuitable for proteomic analysis or protein coverage might be severely compromised. Extraction methods will probably differ for different insect species and they are well described [8–10]. Insect hemolymph usually contains between 20 and 100 mg/ml of total proteins [11]. The exact protein concentration will depend on insect species, development stage, experimental conditions, feeding, and other physiological factors [12]. Because hemolymph tends to coagulate and turns brownish (PO activation) rapidly on exposure to air, for proteomic analysis this can be prevented by mixing immediately with ten volumes of urea denaturing buffer (described in this chapter). For proteomic analysis (label-free quantification), a range from 25 to 50 μg of total proteins is required. Hence, the collection of 1–2 μl of hemolymph is sufficient, and protein concentration determination is not necessarily required. In addition, hemolymph can be processed for individual insects or insects can be pooled, which is useful for those of smaller size.

All solutions should be prepared using ultra-pure water, and all reagents should be of mass spectrometry grade. Several buffers should be freshly made (i.e., between 5 and 30 min before use) unless we indicate otherwise. We advise following all waste disposal regulations when disposing of toxic materials. In addition, it is necessary to follow local regulations for safety procedures with certain reagents indicated throughout the protocols.

3 Materials

This protocol is designed for label-free quantification. For accurate quantitative assessment, it is necessary that there are at least three replications per condition or treatment including the control group. We recommend six replicas per condition or treatment for a better performance. Some proteins may show great variability among individuals. To overcome this issue, several insects (i.e., from three to ten) from the same group can be pooled and considered as a single replica within their treatment group.

3.1 Buffers and Solutions (See Note 1 First)

3.1.1 HEPES Buffer 1 M Solution

1. Prepare 1 M HEPES by adding 2.38 g to 5 ml of distilled water using a magnetic stirrer. Once dissolved, adjust the volume to 10 ml.
2. Make aliquots of 1 ml and store at −20° until use (the solution will remain stable for up to 2 years).

3.1.2 50 mM ABC Buffer (Ammonium Bicarbonate/NH_4HCO_3)

Prepare 50 mM ABC by dissolving 40 mg in 10 ml of water. Use always a freshly made solution. Keep at room temperature until use.

3.1.3 Dithiothreitol (DTT) 1 M Stock Solution (See Note 2)

Weigh 1.54 g of DTT and dissolve in 10 ml of water. Aliquot into 1 ml tubes and store at −20 °C. Stocks are stable for 1 year.

3.1.4 55 mM Iodoacetamide Solution (See Note 3)

Prepare a 55 mM iodoacetamide solution by dissolving 10.2 mg iodoacetamide in 1 ml of ABC buffer. Store in small aliquots at −20 °C (50–100 μl). The solution should be protected from light.

3.1.5 10 mM Dithiothreitol Solution in ABC Buffer (for Protein Digestion)

To make 1 ml of 10 mM DTT in ABC buffer, dilute 10 μl of the 1 M DTT solution in 990 μl ABC buffer. Use always a freshly made solution.

3.1.6 Urea Denaturing Buffer

Composition: 6 M urea, 2 M thiourea, and 10 mM HEPES (pH 8.0).

1. To prepare 100 mm of urea denaturing buffer, dissolve 36 g urea and 15.2 g thiourea in 40 ml water using a magnetic stirrer (although the reaction is highly endothermic (it gets really cold), do not heat).
2. Place the recipient in a water bath at 20 °C to bring the solution back to room temperature.
3. Add 1 ml of the 1 M HEPES stock solution. Adjust the pH to 8 by slowly adding the required amount of 1 N NaOH (1 M). Adjust the final volume to 100 ml.
4. Add 2.5 g of resin beads AG501-X8 Bio-Rex MSZ 501 (Bio-Rad), and stir for 1 h (this step is optional but highly recommended because it removes cyanide, a protein-damaging compound from urea solution). Filter or decant the solution to remove the resin. This solution can be stored at +4 °C for 1 year in a tightly closed flask.

3.1.7 LysC Solution (See Note 4)

Dissolve the LysC protease (sequencing grade) to a final concentration of 0.5 μg/μl in 50 mM ABC buffer. Prepare just before use and keep on ice. Small aliquots can be stored at −20 °C. Avoid freeze-thaw cycles.

3.1.8 Trypsin Solution

Dissolve the trypsin protease (sequencing grade) to a final concentration of 0.5 μg/μl in 50 mM ABC buffer. Prepare just before use and keep on ice. Small aliquots can be stored at −20 °C. Avoid freeze–thaw cycles.

3.1.9 Buffer A∗ (Sample Activation Buffer)

Prepare a fresh solution of 5% acetonitrile and 3% trifluoroacetic acid (TFA). For 50 ml of buffer A∗, add 2.5 ml of acetonitrile and 1.5 ml of TFA (99.9%) to 46 ml of water. Prepare just before use and keep at room temperature.

3.1.10 Buffer A (Tip Equilibration and Washing Buffer)

Prepare a fresh solution of 5% acetonitrile and 0.1% formic acid. For 50 ml of buffer A, add 2.5 ml of acetonitrile and 50 μl of formic acid (99.8%) to 47.5 ml of water. Prepare just before use and keep at room temperature.

3.1.11 Buffer B (Elution Buffer)

Prepare a fresh solution of 80% acetonitrile and 0.1% formic acid. For 10 ml of buffer B, mix 8 ml of acetonitrile, 1.95 ml of water, and 50 μl of formic acid (99.8%) adding them in this order. Prepare just before use and keep at room temperature.

4 Methods

For label-free quantification, a minimum of three samples is advised per group or treatment, but we would recommend at least six.

4.1 Insect Sample Preparation Using the Whole Body

1. Select an appropriately sized mortar and pestle to grind the insects in.
2. Pour liquid nitrogen into the mortar, add the insects, and grind up to a powder. Take care not to allow the nitrogen to completely evaporate, but if it does, allow it only to happen for a few seconds.
3. Quickly transfer the powdered insect body or tissue to a 1.5 ml sample tube standing in liquid nitrogen or a dry ice/ethanol bath using a small metallic scale spoon, which should be pre-chilled by dipping it in liquid nitrogen.
4. In a separate 1.5 ml sample tube, add urea denaturing buffer at a proportion of 1 ml per 100 mg of powdered tissue. The urea solution should be at room temperature, while keeping the remaining powder always cold, ideally on a dry ice box. Preserve the rest of the powdered material at −80 °C. Scale the ratio up or down for a correct solubilization of the material. Mix very well by pipetting and incubate at room temperature for 5 min.
5. Centrifuge the samples at 10,000–12,000 × *g* for 5 min at room temperature. Recover the supernatant without perturbing the pellet of insoluble material.
6. Determine protein concentration by a suitable method (i.e., bicinchoninic acid assay (BCA) or similar).
7. Adjust the protein concentration to 5–10 μg/μl using urea denaturing buffer as a solvent. At this concentration range, only a final volume of 20 μl/sample is required for the experiment.
8. Proceed immediately to Subheading 4.4.

4.2 Insect Sample Preparation Using Hemolymph

1. Extract insect hemolymph using the method of your choice observing our recommendations of general considerations section of this chapter.
2. Immediately after extraction, mix 1 or 2 μl of hemolymph/sample with 20 μl of urea denaturing buffer in a 0.2 ml tube. Mix well by pipetting.
3. Proceed immediately to Subheading 4.4.

4.3 Insect Sample Preparation Using Fatbody

1. Place the insect on a cold surface, for example, a glass Petri dish on ice.
2. Dissect the specimen as quickly as possible, and collect a portion of the fatbody.
3. Transfer a portion of the fatbody tissue to small pre-chilled mortar, and follow the same instructions as given for the whole body, including the proportion 1 ml of urea buffer/100 mg of insect sample.

4.4 Sample Treatment and Digestion Protocol

1. Add 2.5 μl of DTT 10 mM solution (see buffers and solutions section) to each of the tubes containing 20 μl of protein solution in urea denaturing buffer, and incubate for 30 min at room temperature.
2. Add 2.5 μl of 55 mM iodoacetamide solution to each tube, and incubate for 30 min at room temperature in the dark.
3. Add 2 μl of 0.5 μg/μlLysC solution (ratio of ~1 μg of LysC solution/50 μg of protein sample), and incubate for 3 h at room temperature.
4. Dilute samples with four volumes of ABC (final concentration of urea should be below 2 M) by adding 90 μl of ABC buffer.
5. Add 2 μl of 0.5 μg/μl trypsin (ratio of ~1 μg of trypsin solution/50 μg of protein sample or 1 μg trypsin/50 μg sample protein), and incubate overnight at room temperature.
6. The following day (12–16 h after **step 7**), prepare fresh Buffer A* and stop the digestion by acidifying the sample with the required amount of buffer A (normally 10 μl). The pH should be less than 2.5. Test the pH by transferring 1 or 2 μl of digestion solution to pH strips.
7. Proceed with peptide purification following the *StageTip Purification Protocol.* Digested peptides can also be stored at −20 °C until use for short term (*see* **Note 5**).

4.5 StageTip Purification Protocol

This procedure is inspired by previous work intended for micropurification, enrichment, pre-fractionation, and storage of peptides for proteomic analysis using StageTips [13]. We have introduced minor modifications to some reagents and protocol steps. The idea is to use 200 μl pipette tips (filterless) as reverse-phase chromatographic columns. Small disks of a paper-type reverse-phase matrix disk of C18 (3 M™ Empore™ C18 Extraction Disks) are cut and tightly introduced near the end of the tip (*see* Fig. 1). The matrix is then activated and equilibrated, and the sample is applied. This “microcolumn” is washed and the samples are eluted before being applied to the LC-MS machine. All centrifugation steps are carried out at room temperature.

1. Prepare the 200 μl tips for peptide collection upon digestion of the protein sample (so-called StageTips). One tip will be used per sample. Label each tip according to the sample and replica code. Using a puncture, stack four small disks (from 0.4 to 0.6 mm of the 3 M™ Empore™ C18 Extraction Disks) inside the 200 μl tip, and pack them near the narrower extreme of the tip (*see* Fig. 1 for reference).
2. Using an awl, make a hole that has a diameter that is slightly smaller than the broadest part of the pipette tip in the cap of a 2 ml plastic tube (Eppendorf-type). Place a tip containing the

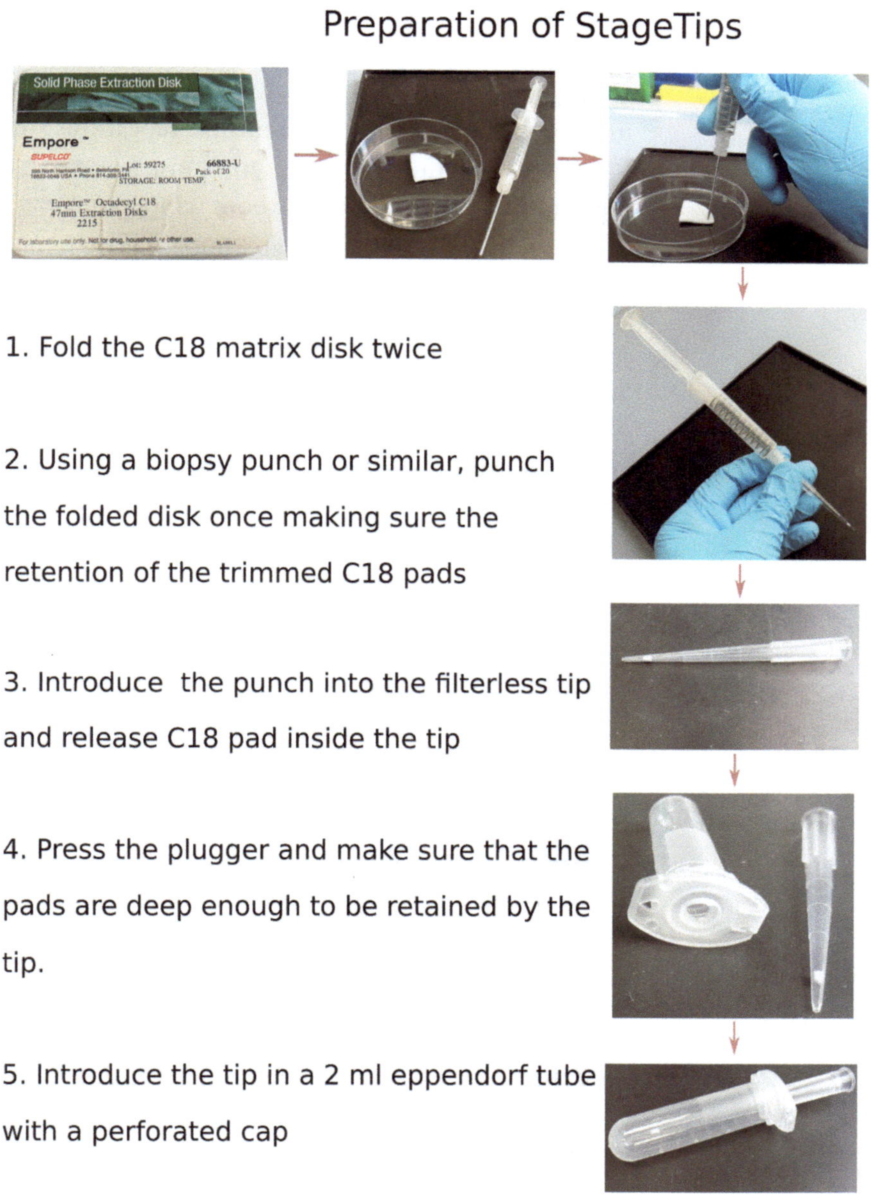

Fig. 1 Photograph of sequential steps to a correct preparation of the StageTips that will be used for micro-purification and enrichment of digested peptides for LC-MS proteomic analysis

paper C18 disks into the 2 ml plastic tube (Eppendorf-type). This tube will be the holder and collecting reservoir for the tips (*see* Fig. 1 for reference).

3. Activate each tip by adding 100 μl of methanol (LC-MS grade).
4. Centrifuge the tip-carrying tubes for 10 s at 10,000 × *g*, and check that some methanol is retained on the matrix disks. If all the solvent goes through, replace the tip with a new one and

repeat the procedure. This is to ensure that the matrix is tightly packed inside the tip.

5. Spin the tip-carrying tubes for 45 s at 3000 × *g* to remove all methanol. If some methanol remains, repeat this step.
6. Equilibrate the tips by adding 200 μl of Buffer A and centrifuge at 3000 × *g* for 5 min at room temperature.
7. Spin the tip-carrying tubes at 3000 × *g* to remove Buffer A. Proceed as soon as possible to the next step to avoid the matrix becoming dry.
8. Add the acidified samples (with Buffer A*; from **step 8**, Subheading 4.4) to their corresponding tips, and centrifuge at 3000 × *g* for 5 min at room temperature.
9. Wash the tips by adding 200 μl of Buffer A and centrifuge at 3000 × *g* for 5 min at room temperature.
10. Place the tip into a tip rack and store at +4 °C until the desired time for elution and analysis. The samples can be stored for up to 3 months under these conditions.
11. For sample elution, prior to LC-MS analysis, add 100 μl of elution Buffer B to each tip, and place them inside the hole practiced on the cap of a 2 ml Eppendorf tube (Fig. 1), using the plastic tube as a holder. The hole should be big enough to tightly place the tip on it but small enough to hold and retain the upper part of the tip.
12. Centrifuge the tubes holding the tips at 3000 × *g* for 5 min at room temperature. If the elution is incomplete, repeat the operation until no liquid is retained inside the tip.
13. Evaporate the solvent using a SpeedVac machine and resuspend the peptides in an appropriate volume of the buffer of choice for the liquid chromatographic step (LC). The samples are ready to be assayed in the LC-MS machine.

5 Notes

1. Use gloves during the whole procedure to avoid sample contamination from the operator's epithelial cells and also to prevent toxicity or irritation produced by some reagents when they come into contact with the skin.
2. The use of DTT is intended to reduce disulfide bonds and help to reduce the complexity of protein structure.
3. Iodoacetamide reacts with reduced disulfide bonds and prevents their re-formation.
4. LysC is non-sensitive to high urea concentrations (up to 8 M) and enhances later digestion with trypsin. Both proteases have the same cleavage specificity.

5. Prolonged storage of digested proteins in plastic tubes may result in irreversible loss of peptides due to binding to the walls of the tubes.

Acknowledgments

ARR and JR were supported by the Collaborative Research Centre (CRC) via project 973 "Priming and Memory of Organismic Responses to Stress," project C5. We are grateful to Arpita Nath and Sophie Armitage from Freie Universität Berlin for the nice comments about this protocol.

References

1. Rolff J, Reynolds S (2009) Insect infection and immunity. Press, Oxford University
2. Hillyer JF (2016) Insect immunology and hematopoiesis. Dev Comp Immunol 58: 102–118
3. Rolff J, Schmid-Hempel P (2016) Perspectives on the evolutionary ecology of arthropod antimicrobial peptides. Philos Trans R Soc Lond B Biol Sci 371:20150297
4. Shishkova E, Hebert AS, Coon JJ (2016) Now, more than ever, proteomics needs better chromatography. Cell Syst 3:321–324
5. Tyanova S, Temu T, Carlson A et al (2015) Visualization of LC-MS/MS proteomics data in MaxQuant. Proteomics 15:1453–1456
6. Sieksmeyer T, He S, Kuropka B, et al (2019) Eating in a losing cause: anorexia and macronutrient manipulation by a cockroach fails to boost immunity, bioRxiv. https://doi.org/10.1101/652826
7. Makarova O, Rodriguez-Rojas A, Eravci M et al (2016) Antimicrobial defence and persistent infection in insects revisited. Philos Trans R Soc Lond B Biol Sci 371:20150296
8. MacMillan HA, Hughson BN (2014) A high-throughput method of hemolymph extraction from adult drosophila without anesthesia. J Insect Physiol 63:27–31
9. Brakefield PM, Beldade P, Zwaan BJ (2009) Hemolymph extraction from various developmental stages of the African butterfly Bicyclus anynana. Cold spring Harb Protoc 2009:pdb.prot5214
10. Borsuk G, Ptaszyńska AA, Olszewski K et al (2017) A new method for quick and easy hemolymph collection from Apidae adults. PLoS One 12:e0170487
11. Feir D, Krzywda L (1969) Concentration of insect hemolymph proteins under various experimental conditions. Comp Biochem Physiol 31:197–204
12. Resh VH, Cardé RT (2009) Encyclopedia of insects. Elsevier/Academic Press, Cambridge
13. Rappsilber J, Mann M, Ishihama Y (2007) Protocol for micro-purification, enrichment, pre-fractionation and storage of peptides for proteomics using StageTips. Nat Protoc 2:1896–1906

Part II

Cellular Response

Chapter 4

Quantification of Blood Cells in *Drosophila melanogaster* and Other Insects

Sean Corcoran and Katja Brückner

Abstract

The invertebrate model organism *Drosophila melanogaster* and other insects have blood cell systems that serve important roles in development, innate immunity, physiology, wound healing, and regeneration. Assessing blood cell numbers is key when investigating these processes. In insects, quantification of blood cells (hemocytes) is often complicated by the presence of sessile, or resident, hemocyte populations that resist bleeding with the hemolymph. Here we describe methods for the absolute quantification of total hemocyte populations of *Drosophila*, in particular macrophage-like plasmatocytes and crystal cells that promote melanization. We describe methods for marking blood cells by genetic reporters or phagocytic labeling followed by their release and semi-automated quantification. In addition, we summarize a method for the quantification of crystal cells, based on melanization. We discuss adaptations of the protocols for other blood cell types and other insect species and extend them for the combined use with cell biological approaches.

Key words *Drosophila melanogaster*, Insect, Hemocyte, Blood cell, Transgenic reporter, Phagocytosis, Macrophage, Plasmatocyte, Crystal cell, Melanization

1 Introduction

Quantification of blood cells is an important task when investigating cellular responses to immune challenges, genetic alterations, injury, physiological changes, and normal and abnormal development. In the model organism *Drosophila melanogaster* and in other insects, the assessment of blood cells (hemocytes) is complicated by the fact that at most developmental stages, large fractions of the hemocyte population are not freely circulating; in the embryo, larval instars, and the adult (imago), hemocytes are often found in sessile, i.e., resident, clusters at internal sites (reviewed in [1–4]).

For example, *Drosophila* larvae harbor resident sites of blood cells from two origins. Hemocytes of the embryonic lineage are organized in segmentally repeated hematopoietic sites, also known as hematopoietic pockets (HPs); some hemocytes also accumulate in clusters at the dorsal vessel, a heart-like organ [1, 2]. Larvae

Federica Sandrelli and Gianluca Tettamanti (eds.), *Immunity in Insects*, Springer Protocols Handbooks,
https://doi.org/10.1007/978-1-0716-0259-1_4,

further possess a hematopoietic organ, the lymph gland (LG), which develops at the anterior end of the dorsal vessel [3, 5]. The fraction of circulating hemocytes increases toward the end of larval development: while in the first and second instar larva essentially all hemocytes are resident, increasing amounts of hemocytes lose adhesion and enter circulation over the course of the third instar larva, culminating in the largely complete release of hemocytes at the onset of metamorphosis [6–12]. In the *Drosophila* adult, hemocytes were recently found to reside in reservoirs along the respiratory epithelia of the head and thorax [4]; to a lesser extent, hemocytes also accumulate at the valves of the heart as has been known in a variety of insect species for decades [13–15].

Many commonly applied methods of hemocyte releases are not quantitative and are rather heavily influenced by variations in hemocyte adhesion. For example, extracting or "bleeding" samples of hemolymph, in the hope it would yield hemocyte counts proportional to the overall number of hemocytes, is very unreliable, as changes in adhesion can mask true changes in total hemocyte numbers. In *Drosophila*, changes in hemocyte adhesion are seen under a variety of conditions. In the larva, besides developmental changes in hemocyte adhesion (*see* above), immune challenges such as wasp infestation result in the release of hemocytes from their resident sites in the HPs and disintegration of the LG [3, 5, 8]. Likewise, manipulation of signaling through oncogene overexpression dramatically reduces hemocyte adhesion [16, 17]. Physiological processes affect hemocyte adhesion and thereby localization, such as starvation driving hemocyte accumulation in the fat body [18]. Mechanical manipulation of larvae also dislodges resident hemocytes, thereby leading to an inadvertent increase in circulating hemocytes [9, 11].

Drosophila and other insects possess myeloid blood cell systems, while lymphoid cells typically are not present in insects. The most abundant blood cell types are macrophage-like cells, in *Drosophila* called plasmatocytes [1–3, 5, 19]. To a lesser extent, crystal cells, an insect-specific immune cell type specialized in melanization, are also present at most developmental stages [1–3, 5, 19]. Lamellocytes are a large immune cell type with roles in parasite encapsulation and are induced to differentiate mainly under immune challenges or stress conditions in the larva [3, 5, 7, 20]. In addition, significant amounts of undifferentiated hemocyte progenitors (prohemocytes) are present in the early embryo and in the larval lymph gland [3, 5].

Here we describe methods for the quantification of *Drosophila* plasmatocytes and crystal cells. In addition, in the Notes section, we also suggest modifications that will allow adaptation of the protocols for other blood cell types and other insect species, and the combined use with assays for cellular responses.

2 Materials

2.1 Reporter Labeling of Hemocyte Populations

The most common way to label hemocyte populations in *Drosophila* relies on transgenic reporters that express fluorescent proteins in hemocyte populations of interest. Reporters can be direct or consist of a GAL4 transgenic or insertion line combined with a fluorescent protein transgenic line as readout (Fig. 1a).

2.1.1 Materials Needed

Examples of commonly used *Drosophila* hemocyte reporter lines are listed below. Typically, a reporter for one immune cell type or reporters for two hemocyte types (marked by two distinct fluorescent proteins) are chosen. In cases where a binary system such as GAL4-UAS is used, both lines are either genetically (re-) combined or are brought together in a F1 genetic cross (Fig. 1a). Lines listed below label predominantly, albeit in some cases not exclusively, the following blood cell types.

1. Plasmatocytes.
   (a) *HmlΔ-DsRed* [9].
   (b) *HmlΔ-GAL4, UAS-GFP* [21].
   (c) *Pxn-GAL4, UAS-GFP* [22].
   (d) *eater-GAL4, UAS-2XEYFP* [23].
2. Crystal cells.
   (a) *BcF6-mCherry* [23].
   (b) *BcF2-GFP* [24].
   (c) *lz-GAL4, UAS-GFP* [25].
3. Lamellocytes.
   (a) *msmF9-mCherry* [23].
   (b) *msmF9-GFP* [26].

2.2 Phagocytic Labeling of Hemocytes

As an alternative to marking hemocytes with a genetic reporter, *Drosophila* plasmatocytes or other phagocytic immune cells (names may vary in other insect species) and their derivatives can be marked by a phagocytic label. Here we use fluorescently labeled bacteria that are injected into the animal and are taken up by plasmatocytes (Fig. 1c).

2.2.1 Materials Needed

1. *Drosophila* larvae.
2. Paintbrush (liner).
3. Spatula.
4. Petri dish, (60 mm).
5. PBS (phosphate buffered saline).
6. Squirt bottle with water.

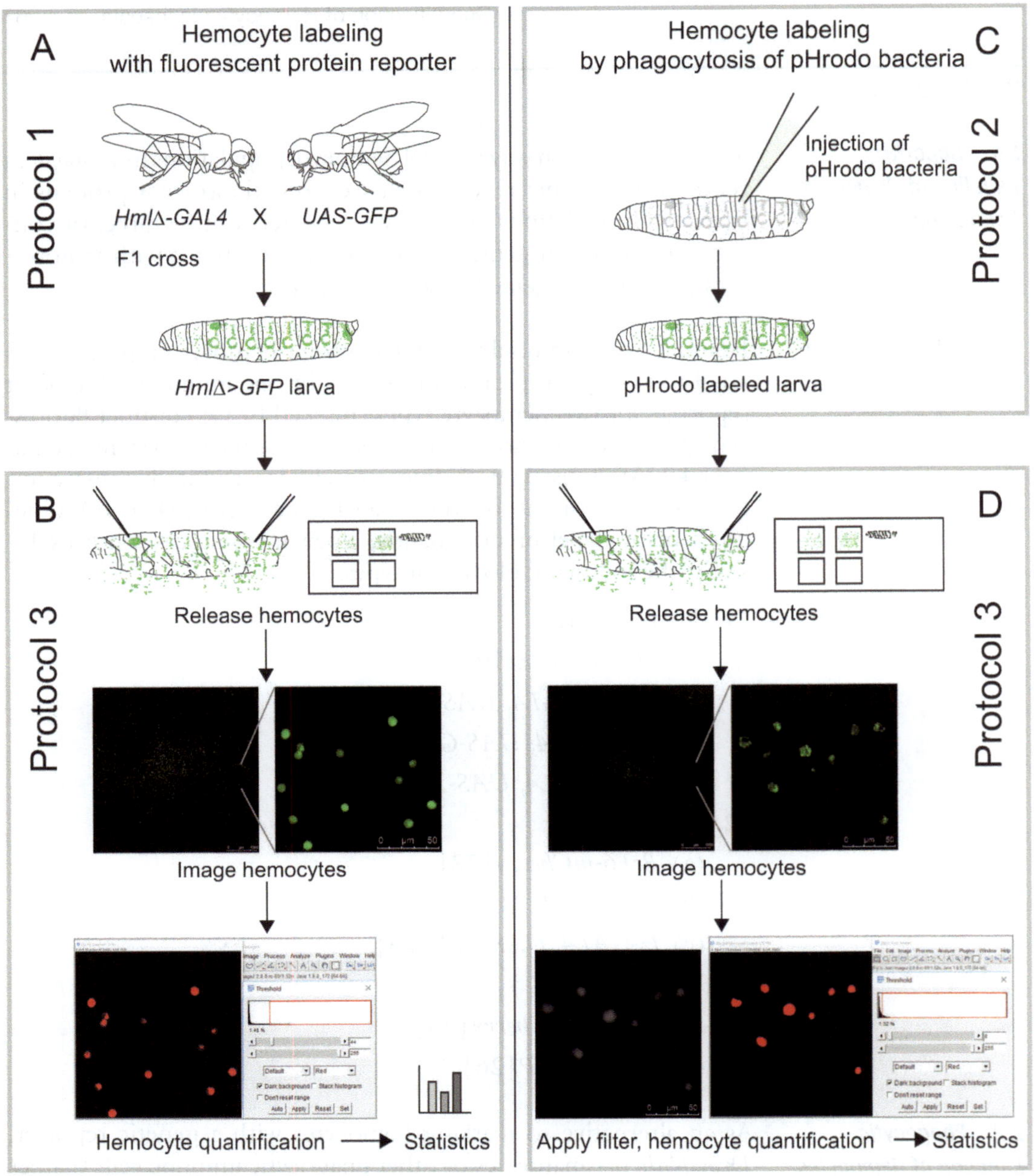

Fig. 1 Quantification of hemocytes by fluorescent reporter or phagocytic labeling. (**a**) Reporter labeling of hemocytes. Example illustrates the progeny of an F1 cross of a hemocyte GAL4 driver (*HmlΔ-GAL4*, mainly expressing in macrophage-like plasmatocytes) and a UAS line of green fluorescent protein (*UAS-GFP*); F1 genotype is *HmlΔ-GAL4/UAS-GFP*. Scheme corresponds to Protocol 1. (**b**) Quantification of hemocytes following (**a**). Release of all hemocytes from a *Drosophila* larva (upper left) and arrangement on the slide (upper right); tile-scan imaging, overview (middle left) and close up image (middle right), showing hemocytes in green; cell counting using particle quantification in ImageJ/Fiji or similar image analysis software (particle counting in red). Data of multiple larvae and experimental conditions are subjected to statistical analysis. Scheme corresponds to Protocol 3. (**c**) Labeling of phagocytic hemocytes (*Drosophila* plasmatocytes) by injection of pHrodo bacteria. Scheme corresponds to Protocol 2. (**d**) Quantification of hemocytes following (**c**). Release of all hemocytes from a *Drosophila* larva (upper left) and arrangement on the slide (upper right); tile-scan imaging, overview (middle left) and close up image (middle right), showing incorporated pHrodo bacteria in phagocytic vesicles in green; the image is processed in ImageJ/Fiji or similar image analysis software by applying a filter (Gaussian Blur) to fuse the signal of phagocytic inclusions within each cell area (lower left, signal in white), followed by cell counting using particle quantification (lower right, particle counting in red). Data of multiple larvae and experimental conditions are subjected to statistical analysis. Scheme corresponds to Protocol 3

7. p200 pipette.
8. pHrodo *E. coli* BioParticles (Invitrogen) (10 mg/ml PBS), here pHrodo Green.
9. Injector (Nanoject II, Drummond).
10. Pulled glass capillaries (capillaries Drummond; puller Sutter Instruments).
11. Forceps (Dumont #5, Fine Science Tools).
12. Silicone dissection plate (Petri dish filled with Sylgard (Dow Corning) stained with charcoal powder (Sigma), cured).
13. Stereo microscope.
14. *Drosophila* food source.

2.3 Quantification of Hemocytes Marked by Reporters or Phagocytic Label

Once hemocytes are labeled by one of the above methods, all hemocytes of the animal are released under a stereoscope, imaged, and quantified using particle counting software (Fig. 1b, d).

2.3.1 Materials Needed

1. *Drosophila* larvae labeled according to Subheadings 2.1 or 2.2.
2. Paintbrush (liner).
3. Petri dish, 60 mm.
4. PBS.
5. Squirt bottle with water.
6. p200 pipette.
7. Optional: metal block on ice in containment box.
8. Glass slide.
9. Hydrophobic PAP pen (Fisher Scientific).
10. Dissection tools: forceps (Dumont #5, Fine Science Tools), two dissection needles (or hypodermic needles mounted on small (e.g., 1 ml) syringes) (Fisher).
11. Humid chamber: box with lid, walls inside lined with wet paper towel (if doing more than one slide of released hemocytes).
12. Fluorescence stereo microscope (e.g., Leica M205).
13. Inverted fluorescence microscope (e.g., Leica DMI4000) with camera and image acquisition software; tilescan capability is preferred but not a strict requirement.
14. Computer.
15. ImageJ/Fiji (NIH Image [27]) or other software with particle counting capabilities (e.g., MetaMorph (Molecular Devices)).

2.4 Quantification of Crystal Cells Marked by Melanization

Crystal cells express prophenoloxidase, which bestows them with melanization potential. This classic protocol uses heat to induce spontaneous melanization that results in visible blackening of crystal cells, as previously described by M.T.M. Rizki [28].

Quantification is performed by manual counting of blackened crystal cells on the heat denatured, yet unfixed material (Fig. 2a).

2.4.1 Materials Needed

1. *Drosophila* larvae of genotype and age of interest.
2. Paintbrush (liner).
3. Spatula.
4. Petri dish, 60 mm.
5. p200 pipette.
6. PBS.
7. Squirt bottle with water.
8. Metal block on ice in containment box.
9. Glass slide.
10. 1.5 ml microcentrifuge tubes.
11. Heating block set at 65 °C.
12. Thermometer.
13. Stereo microscope or other microscope with light source for crystal cell counting, preferred magnification 100–150×.

3 Methods

3.1 Reporter Labeling of Hemocyte Populations

Set up genetic cross of GAL4 driver and fluorescent protein UAS transgenic line, or use existing fluorescent reporter line. Collect F1 offspring after flipping the cross or stock on new food in time intervals of choice.

3.2 Phagocytic Labeling of Hemocytes

1. Set up Drummond Nanoject with a sharp glass capillary.
2. Fill capillary with Invitrogen *E. coli* pHrodo BioParticles (10 mg/ml) in PBS.
3. Isolate larvae by scooping some fly food from the vial with larvae using a spatula, and rinsing larvae with water from a squirt bottle in a Petri dish. Recommended larval age is early third instar (86 h after egg laying (AEL) or older).
4. Using a paintbrush, transfer larvae onto a silicone dissection plate.
5. Optional: Inspect larvae for age/size/genotype using a cold block and dissection microscope, and choose desired larvae. This step may be dispensable when applying short time windows of egg laying to control for larval age.
6. To inject larvae under the dissection microscope, use one hand to stabilize a larva with forceps and with the other use the sharp glass capillary to penetrate the larva in the dorsal side at around a 30° angle. Inject 69 nl of pHrodo BioParticles into the larva.

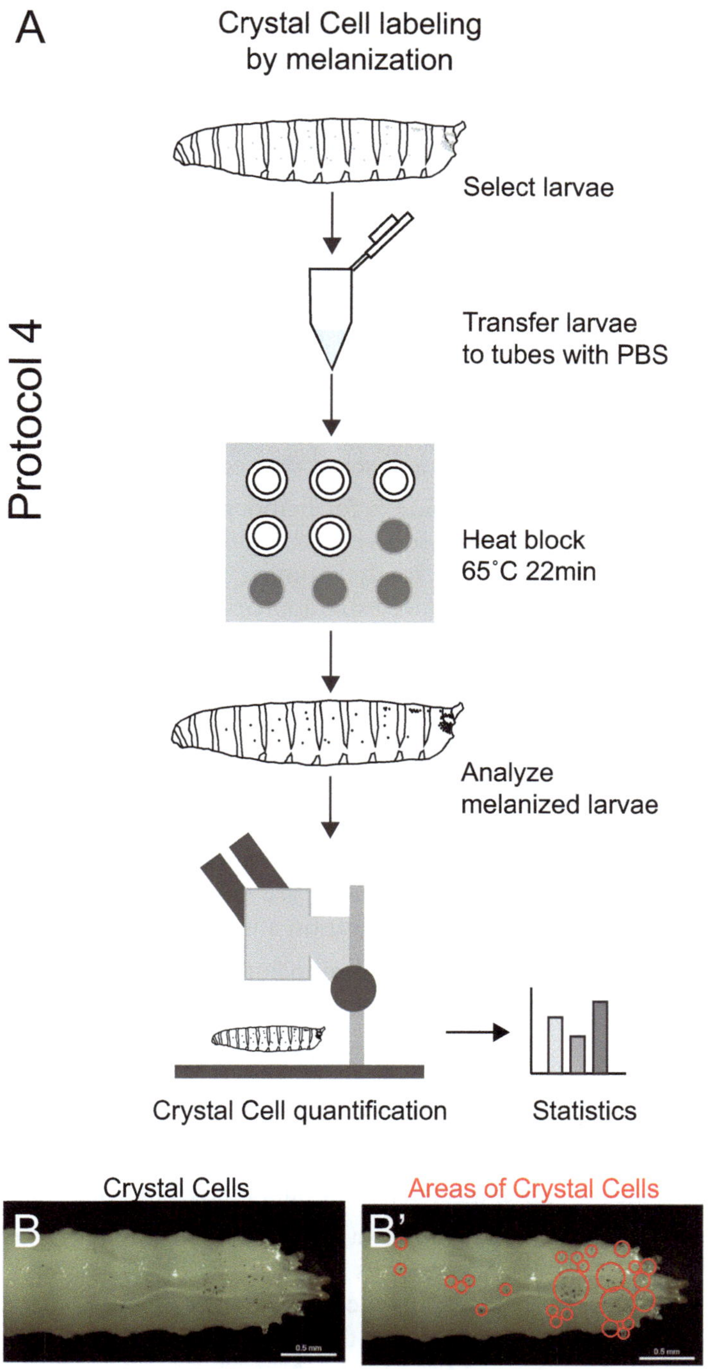

Fig. 2 Quantification of crystal cells by heat-induced melanization. (**a**) Crystal cell melanization. *Drosophila* larvae are selected, placed in microcentrifuge tubes with PBS, and heated to 65 °C for 22 min in a heat block. This results in the melanization (blackening) of crystal cells. Larvae are then removed from the tubes, and black cells are manually counted under a dissecting microscope. Data of multiple larvae and experimental conditions are subjected to statistical analysis. Scheme corresponds to Protocol 4. (**b**) *Drosophila* larva with melanized crystal cells, dorsal view of posterior area. (**b′**) Same image as in (**b**), with locations of crystal cells marked by red circles

7. Use paintbrush to place injected larva into vial of food.
8. Repeat **steps 4–7** with desired number of larvae.
9. Incubate injected larvae for 4 h at 25 °C.

3.3 Quantification of Hemocytes Marked by Reporters or Phagocytic Labeling

1. In a containment box, place metal block and add ice around it to cool it down. Place one glass slide on the metal block.
2. Isolate larvae by scooping some fly food from the vial with larvae using a spatula, and rinsing larvae with water from a squirt bottle in a Petri dish.
3. Using a paintbrush, transfer larvae to the glass slide on the cold metal block.
4. Under a fluorescence stereo microscope (M205 Leica), line up the larvae and select desired age/size/genotype.
5. Prepare slide/s for hemocyte release (*see* **Note 1**). With PAP pen draw square/rectangular wells on clean glass slide corresponding to the number of larvae (max wells on a single slide = 8), well size approximately 3–4 mm.

   When a microscope without tilescan function is used, wells should be small enough to be imaged in single microscope field (e.g., 2 mm); two small wells per larva are recommended.
6. Fill each well with 10–30 μl of PBS but avoid overflowing.
7. Place slide with PAP pen wells under a fluorescence dissecting microscope at room temperature (no cold metal block). Hemocyte releases are done under fluorescence to allow efficient identification and release of all hemocytes.
8. Use needle or clean paintbrush to lift one larva from the cold block into a PAP pen well of the slide.
9. Slowly agitate larva/perturb resident hemocytes by gently rolling or tapping larva so that most hemocytes are circulating.
10. Hold down larva with one needle and using another needle to poke holes in the posterior and anterior areas of the larva. Hemocytes should begin to flow out of these holes into the PBS.
11. Scrape/tap larva at various locations to remove as many resident hemocytes as possible. If you do not want to release hemocytes of the LG, locate the LG, hold down the larva with a needle close to the LG, and carefully release hemocytes around the area (*see* **Note 2**).
12. Once almost all hemocytes are released, place the carcass on the side of well.
13. Repeat **steps 8–12** for as many larva as you want to release.
14. Let hemocytes settle to the bottom of each well until all cells are in the same focal plane (may take between 10 and 20 min).

15. During this time, count any remainder of hemocytes in the carcasses (usually 5–30 hemocytes) and record the numbers; they will later be added to the total sum of hemocytes.
16. Carefully place slide/s in a moist chamber and ensure wells are not drying out. If needed, add some more PBS.
17. Place slide on an inverted fluorescence microscope.
18. Using the tilescan function of the microscope, image the entire well and repeat for all wells. If a microscope without tilescan function is used, image each entire PAP pen well.
19. Save and export images.
20. Open images in ImageJ/Fiji software to quantify fluorescent hemocytes in each well.
21. ImageJ/Fiji Protocol:
    (a) Open file on ImageJ/Fiji.
    (b) Convert image type to 8bit.
    (c) If cells are labeled by phagocytic particles, a filter is applied first before cells are counted. Click *Process* and select *Filter.* Then choose *Gaussian Blur* and set the *Sigma* value high enough so that multiple particles found in each hemocyte blur together into one continuous area.
    (d) To count cells labeled by either method, click *Adjust* and select *Threshold.* Select *Dark Background* and increase lower threshold bar until every grayscale cell becomes red.
    (e) When all cells are accounted for, select *Analyze* and *Analyze Particle.* Click box for *Summarize.* Optionally, set size and circularity for cells of choice. Click *Ok* and observe total number of cells counted.
    (f) Where cells are clumped together and hard to resolve at the individual cell level with threshold, use a hand counter to manually count for adjustment.
22. Combine carcass count, ImageJ/Fiji counts of the two wells per larva, and manual adjustment count to calculate the total hemocyte number per larva.

3.4 Quantification of Crystal Cells Marked by Melanization

1. Set heating block to 65 °C (thermometer reading).
2. Add 250 μl PBS into microcentrifuge tubes; prepare 1 tube per larva to be analyzed. Alternatively, melanization can be performed in small pools of larvae of up to 5 animals, if no specifics of each larva need to be recorded.
3. Isolate larvae by scooping some fly food from the vial with larvae using a spatula, and rinsing larvae with water from a squirt bottle in a Petri dish. Recommended larval age is late second or third instar.

4. Optional: Place larvae on glass slide on metal block cooled by ice in containment. Inspect larvae for age/size/genotype using a cold block and dissection microscope, and choose desired larvae. Record specifics of each larva. This step may be dispensable when applying short time windows of egg laying to control for larval age.
5. Transfer 1 larva each into the prepared tubes with PBS, or set up pools of larvae (*see* above). Proceed swiftly (*see* **Note 3**).
6. Place tubes in heating block at 65 °C for exactly 22 min.
7. Remove larvae and place on slide.
8. Under a stereoscope with brightfield, count blackened cells:
   (a) Using a paintbrush and high magnification (100×–200×), count crystal cells by using a hand counter.
   (b) As needed, turn the focus knob of the microscope and count black cells in each focal plane; avoid double counting cells.
   (c) A recommended counting scheme is as follows:
   - Position larva dorsal up and begin counting black cells on dorsal side.
   - Use paintbrush to gently roll/hold larva on the side and count black cells on side.
   - Continue rolling the larva and counting each black cell on ventral side and other lateral side.
   - Lastly, use a needle to poke the larva near the posterior end so that the caudal end is pointing up; readjust the focus for counting caudal crystal cells.
9. Record number of crystal cells.
10. Repeat for the rest of the larvae (*see* **Note 4**).

4 Notes

1. This step can be completed ahead of time, which helps with experiment set-up.
2. This is only relevant in late second and third instar larvae, when hemocytes switch on differentiation markers and become capable of phagocytosing pHrodo bacteria.
3. Extended exposure in PBS (or water) should be avoided, as this may reduce melanization efficiency; as reference, controls should be processed in parallel with all experiments.
4. Since heat-denatured larvae are not fixed, process all larvae for counting within ~15 min to avoid decomposition. If needed, perform heat treatment in smaller groups of larvae.

5. This chapter describes protocols for the quantification of hemocytes in *Drosophila* and other insects. The methods work independently of the behavior of hemocytes to adhere to resident, or "hidden," sites within the organism, which would be resistant to hemolymph sampling or" bleeding." Since our methods are quantitative, their use allows to eliminate variation across platforms, experimenters, and laboratories, which currently makes comparison of hemocyte datasets between studies and labs very difficult. The main method presented in this chapter is semi-automated and therefore yields very reproducible results and reduces the time needed for quantification.
6. Our protocols can be adapted to a variety of blood cell types, developmental stages, conditions, and organisms. For example, adaptation to other hemocyte populations, such as lamellocytes, mainly requires a specific fluorescent reporter (*see* Subheading 2.1) or can be achieved by simple staining, e.g., with phalloidin, which particularly highlights the large dimensions and unique morphology of lamellocytes.
7. Regarding different developmental stages, the use of fluorescent reporters or labeling with fluorescent pHrodo bacteria is ideal also for hemocyte quantification in adult *Drosophila*. Here the only important precaution is to thoroughly scrape hemocytes from all internal sites of the fly, as hemocytes accumulate in hidden reservoirs lining the respiratory epithelia of the head and thorax [4] (In press now).
8. Hemocyte release methods are also very useful for the quantification of resident versus circulating hemocyte populations. A video article on the use of a very similar hemocyte release method has been published previously [11].
9. The protocols can be equally useful for research in other insect species (e.g., other *Drosophilidae*, or more distant insect species), even in cases where transgenic fluorescent reporters are not available. For example, as long as animal sizes are suitable for injection, labeling based on phagocytosis of fluorescent material (such as pHrodo bacteria),is feasible and may simply need adjustment of injected volumes and incubation times. As in *Drosophila*, this labeling approach would be restricted to phagocytic cells and their derivatives. Likewise, melanization of PPO containing crystal cells and labeling of lamellocyte-like cells with the help of phalloidin staining are approaches that may be applied for the quantification of specific hemocyte types in other insect species.
10. Lastly, our approaches of quantitative releases of hemocytes can easily be combined with a variety of methods that visualize other markers (e.g., antibody staining, fluorescent in situ/FISH approaches, etc.), or assess cell biological readouts

such as apoptosis assays, or BrdU/EdU incorporation assays. EdU incorporation assays have been performed on hemocytes ex vivo, after the release of hemocytes into cell culture medium [9].

11. In all these cases where further treatment of hemocytes is desired, the only cautionary point is that methods requiring fixation or washes of hemocytes may inadvertently lead to some loss of hemocytes from the wells, if hemocytes may not completely adhere to the surface of the glass slide or cell culture dish. With the basic protocols presented in this chapter, cells that do not attach are still accounted for, which in some cases may make an important difference for the exact quantification of hemocyte populations.

Acknowledgments

We thank S. Sinenko and B. Mathey-Prevot for the transgenic line *HmlΔ-GAL4* and members of the Brückner lab for feedback on the protocols. This work was supported by grants from the National Science Foundation 1326268, National Institutes of Health 1R01GM112083 and 1R01GM131094, and the American Cancer Society RSG DDC-122595 (to KB).

References

1. Gold KS, Brückner K (2014) *Drosophila* as a model for the two myeloid blood cell systems in vertebrates. Exp Hematol 42:717–727
2. Gold KS, Brückner K (2015) Macrophages and cellular immunity in *Drosophila* melanogaster. Semin Immunol 27:357–368
3. Banerjee U, Girard JR, Goins LM, Spratford CM (2019) *Drosophila* as a genetic model for hematopoiesis. Genetics 211:367–417
4. Sanchez Bosch P, Makhijani K, Herboso L, Gold KS, Baginsky R, Woodcock KJ, Alexander B, Kukar K, Corcoran S, Jacobs T, Ouyang D, Wong C, Ramond EJV, Rhiner C, Moreno E, Lemaitre B, Geissmann F, and Brückner K (2019) Adult *Drosophila* lack hematopoiesis, but are a model for organismal immunity centered at the blood reservoir of the respiratory epithelia. Dev Cell https://doi.org/10.1016/j.devcel.2019.10.017
5. Letourneau M, Lapraz F, Sharma A, Vanzo N, Waltzer L, Crozatier M (2016) *Drosophila* hematopoiesis under normal conditions and in response to immune stress. FEBS Lett 590:4034–4051
6. Lanot R, Zachary D, Holder F, Meister M (2001) Postembryonic hematopoiesis in *Drosophila*. Dev Biol 230:243–257
7. Markus R, Laurinyecz B, Kurucz E, Honti V, Bajusz I, Sipos B, Somogyi K, Kronhamn J, Hultmark D, Ando I (2009) Sessile hemocytes as a hematopoietic compartment in *Drosophila* melanogaster. Proc Natl Acad Sci U S A 106: 4805–4809
8. Grigorian M, Mandal L, Hartenstein V (2011) Hematopoiesis at the onset of metamorphosis: terminal differentiation and dissociation of the *Drosophila* lymph gland. Dev Genes Evol 221:121–131
9. Makhijani K, Alexander B, Tanaka T, Rulifson E, Brückner K (2011) The peripheral nervous system supports blood cell homing and survival in the *Drosophila* larva. Development 138:5379–5391
10. Makhijani K, Brückner K (2012) Of blood cells and the nervous system: hematopoiesis in the *Drosophila* larva. Fly (Austin) 6:254–260
11. Petraki S, Alexander B, Brückner K (2015) Assaying blood cell populations of the *Drosophila*

melanogaster larva. J Vis Exp (105):e52733. https://doi.org/10.3791/52733

12. Makhijani K, Alexander B, Rao D, Petraki S, Herboso L, Kukar K, Batool I, Wachner S, Gold KS, Wong C et al (2017) Regulation of *Drosophila* hematopoietic sites by Activin-beta from active sensory neurons. Nat Commun 8:15990
13. Gupta AP (2009) Insect hemocytes. Cambridge University Press, Cambridge
14. Elrod-Erickson M, Mishra S, Schneider D (2000) Interactions between the cellular and humoral immune responses in *Drosophila*. Curr Biol 10:781–784
15. Dionne MS, Ghori N, Schneider DS (2003) *Drosophila* melanogaster is a genetically tractable model host for *Mycobacterium marinum*. Infect Immun 71:3540–3550
16. Asha H, Nagy I, Kovacs G, Stetson D, Ando I, Dearolf CR (2003) Analysis of Ras-induced overproliferation in *Drosophila* hemocytes. Genetics 163:203–215
17. Zettervall CJ, Anderl I, Williams MJ, Palmer R, Kurucz E, Ando I, Hultmark D (2004) A directed screen for genes involved in *Drosophila* blood cell activation. Proc Natl Acad Sci U S A 101:14192–14197
18. Shim J, Mukherjee T, Banerjee U (2012) Direct sensing of systemic and nutritional signals by haematopoietic progenitors in *Drosophila*. Nat Cell Biol 14:394–400
19. Rizki TM (1978) The circulatory system and associated cells and tissues. In: Ashburner M, Wright TRF (eds) The genetics and biology of *Drosophila*, vol 2b. Academic Press, New York, pp 397–452
20. Rizki TM, Rizki RM (1992) Lamellocyte differentiation in *Drosophila* larvae parasitized by Leptopilina. Dev Comp Immunol 16:103–110
21. Sinenko SA, Mathey-Prevot B (2004) Increased expression of *Drosophila* tetraspanin, Tsp68C, suppresses the abnormal proliferation of ytr-deficient and Ras/Raf-activated hemocytes. Oncogene 23:9120–9128
22. Stramer B, Wood W, Galko MJ, Redd MJ, Jacinto A, Parkhurst SM, Martin P (2005) Live imaging of wound inflammation in *Drosophila* embryos reveals key roles for small GTPases during in vivo cell migration. J Cell Biol 168:567–573
23. Tokusumi T, Shoue DA, Tokusumi Y, Stoller JR, Schulz RA (2009) New hemocyte-specific enhancer-reporter transgenes for the analysis of hematopoiesis in *Drosophila*. Genesis 47:771–774
24. Gajewski KM, Sorrentino RP, Lee JH, Zhang Q, Russell M, Schulz RA (2007) Identification of a crystal cell-specific enhancer of the black cells prophenoloxidase gene in *Drosophila*. Genesis 45:200–207
25. Lebestky T, Chang T, Hartenstein V, Banerjee U (2000) Specification of *Drosophila* hematopoietic lineage by conserved transcription factors. Science 288:146–149
26. Tokusumi T, Sorrentino RP, Russell M, Ferrarese R, Govind S, Schulz RA (2009) Characterization of a lamellocyte transcriptional enhancer located within the misshapen gene of *Drosophila* melanogaster. PLoS One 4:e6429
27. Schindelin J, Arganda-Carreras I, Frise E, Kaynig V, Longair M, Pietzsch T, Preibisch S, Rueden C, Saalfeld S, Schmid B, Tinevez J-Y, White DJ, Hartenstein V, Eliceiri K, Tomancak P, Cardona A (2012) Fiji: an open-source platform for biological-image analysis. Nat Methods 9(7):676–682
28. Rizki M (1957) Alterations in the hemocyte population of *Drosophila* melanogaster. J Morphol 100:437–458

Chapter 5

Methods to Quantify In Vivo Phagocytic Uptake and Opsonization of Live or Killed Microbes in *Drosophila melanogaster*

Samuel Liégeois, Wenhui Wang, and Dominique Ferrandon

Abstract

Here we describe different phagocytosis assays in *Drosophila*, using various killed or live microbes (bacteria and fungi). Different ex vivo and in vivo approaches are shown, to quantify larval and adult phagocytosis of microorganisms by hemocytes. We also explain how to perform an in vivo opsonization assay. Altogether, these protocols represent a useful range of tools to the researcher interested in the detailed analysis of phagocytosis in the context of the study of host-pathogen relationships.

Key words Opsonization, Phagocytosis, *Drosophila*, Larva, Hemocytes, Bacteria, Fungi, Yeast, Infection, pHrodo

1 Introduction

The insect immune system comprises several arms, including a systemic humoral response, the melanization response triggered by proteolytic cascades, and the cellular immune response [1]. While much emphasis has been placed on the study of the humoral immune response in the past 30 years because of its efficacy against most microbes, the cellular immune response is much less understood in molecular terms. In *Drosophila*, blood cells are called hemocytes, and most of them are phagocytotic macrophage-like cells named plasmatocytes.

Two waves of hematopoiesis occur during *Drosophila* development. The embryonic/larval lineage originates from the head mesoderm of the embryo, differentiates in the embryo, and subsequently expands in the larva. The progenitor-based lymph gland lineage originates in the dorsal mesoderm of the embryo and differentiates in the late larva [2]. Hemocytes of both lineages persist through pupal development into the adult. During the larval instars, hemocytes derived from embryogenesis spread throughout

Federica Sandrelli and Gianluca Tettamanti (eds.), *Immunity in Insects*, Springer Protocols Handbooks,
https://doi.org/10.1007/978-1-0716-0259-1_5,

the animal and are found in two locations: a subset of them circulates in the hemolymph, and the rest are attached to the body wall in sessile pools. At the end of the third instar larval stage, called the wandering stage, ecdysone signaling induces dispersal and activation of sessile hemocytes upon pupariation, and this facilitates tissue remodeling during metamorphosis. Several hundred blood cells are made in the embryo. This number expands through the larval stages to more than 5000 hemocytes during the pupal stage. At this point, there is a high demand for blood cells to accommodate the extensive histolysis and tissue remodeling occurring during metamorphosis. The total number of hemocytes in the adult ranges between 1000 and 2000 cells per larva [3].

Functionally, phagocytosis does not seem to play a preponderant role in the host defense against Gram-negative bacteria in systemic infection models. It, however, plays an essential role in controlling bacteria that evade the digestive tract [4–6]. As regards Gram-positive bacteria and fungi, phagocytosis fulfills a more important function in host defense as its ablation leads to an enhanced sensitivity to these systemic infections [7, 8]. It may also represent a remaining protection when the humoral immune response is deficient. Investigators initially relied on semiquantitative assays or ex vivo quantitative assays [9]. A common feature of these studies is that they used fluorescently labeled killed microbes, usually *Staphylococcus aureus* as a Gram-positive bacterium, *Escherichia coli* as a Gram-negative bacterium, and zymosan particles to mimic fungal cells. The fluorescence of noninternalized bacteria was then quenched by adding Trypan blue, thereby allowing the visualization of engulfed bacteria only [4, 10, 11] (*see* **Note 10**). A positive control consisted in saturating the phagocytic apparatus by the prior injection of "latex" beads (actually often polystyrene beads), a strategy that can also be used for other insects [10, 12, 13]. A second possible control would have been to use hemocyte-depleted flies [14, 15]. A major difficulty for establishing quantitative assays in adults is the low number of plasmatocytes that can be retrieved, as most of them are sessile. In contrast, it is easy to obtain hundreds of hemocytes by just bleeding a single larva. As it is difficult to inject *Drosophila* larvae, ex vivo assays have been favored. One advantage of using killed microbes is that one can measure the host response in the absence of active interference from the pathogen. However, microbes have developed several strategies to elude phagocytosis, either by blocking it directly or by avoiding detection. Thus, when studying host-pathogen relationships, it is critical to also monitor the uptake of live microorganisms.

In this chapter, we provide descriptions of several techniques currently used in the field to quantify phagocytosis that we routinely use in the laboratory [8, 16]. We selected as examples infection models currently under investigation in the team (the Gram-negative bacteria *Serratia marcescens* and *Pseudomonas aeru-*

ginosa, the mold *Metarhizium anisopliae*, and the yeast *Candida glabrata*), but the protocols described here can be generalized to other microbes.

We shall describe on the one hand methods to quantify microorganism internalization using either ex vivo or in vivo assays and on the other hand a method to study the coating of a particle with proteins that facilitate phagocytosis of the particle by macrophages, a process called opsonization. These different techniques partially overlap and we have therefore decided to detail the common steps in the form of modules (Fig. 1), detailed in Subheading 3. First, development of antibodies (Subheading 3.1) and preparation of microorganisms (Subheading 3.2) are described. Then, collection of hemocytes and preparation of tissues (Subheading 3.3) can be done at the beginning of ex vivo phagocytosis assays (Subheading 3.4) or after incubation with microbes for in vivo assays (Subheading 3.5) and opsonization assays (Subheading 3.6). Of note, the in vivo technique necessitates the injection of microorganisms, a procedure that is straightforward in adults yet challenging in *Drosophila* larvae (reported in Subheading 3.5). The opsonization assay involves a more complex procedure (Subheading 3.6, Fig. 1b). At the end of the different assays, when using non-labelled microbes, the in/out differential immunostaining procedure can be performed to reveal whether the microorganisms have been internalized (Subheading 3.7). It is not the case for one of the ex vivo phagocytosis assays that relies on pHrodo-labeled microorganisms. Finally, the last step described is the sample analysis by microscopy (Subheading 3.8 and Fig. 2).

These methods allow studying phagocytosis not only from the standpoint of the host but within the more meaningful context of host-pathogen relationships in which pathogens attempt to elude or neutralize the host immune response [6, 16–18].

2 Materials

2.1 Drosophila and Microbial Strains: Culture Conditions

1. Two-to-seven-day-old *Drosophila melanogaster* females from the following genotypes: wild-type w^{A5001}, homozygous recombinant *HMLdelta-GAL4,UAS-GFP* flies constitutively expressing GFP in hemocytes [19] and *Tep4* mutants (Bloomington stock #15936), checked for the absence of known contaminants [20, 21], and conventionally reared at 25 °C.
2. *Serratia marcescens* wild-type Db11; overnight culture in LB medium at 37 °C.
3. *Pseudomonas aeruginosa* wild-type PA14; overnight culture in BHB medium at 37 °C.

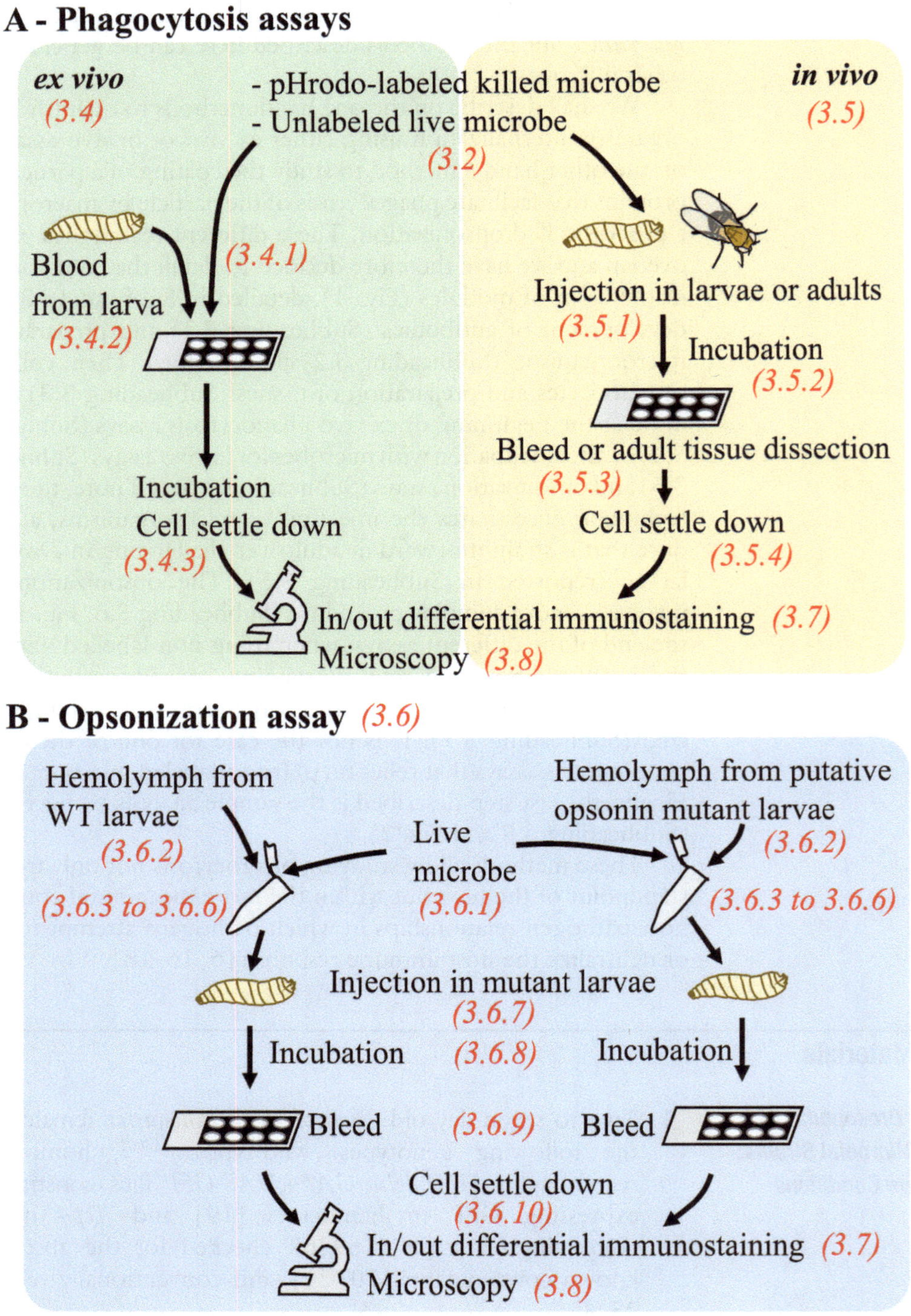

Fig. 1 Overview of the different phagocytosis and opsonization assays. (**a**) phagocytosis assays. This scheme shows the main steps that allow quantifying phagocytosis ex vivo (left) and in vivo (right). Numbers in red indicate the corresponding section in the text. (**b**) Opsonization assays. This scheme shows the main steps of the opsonization assay. Indications in red allow finding the corresponding section in the text to read the full description of the procedure

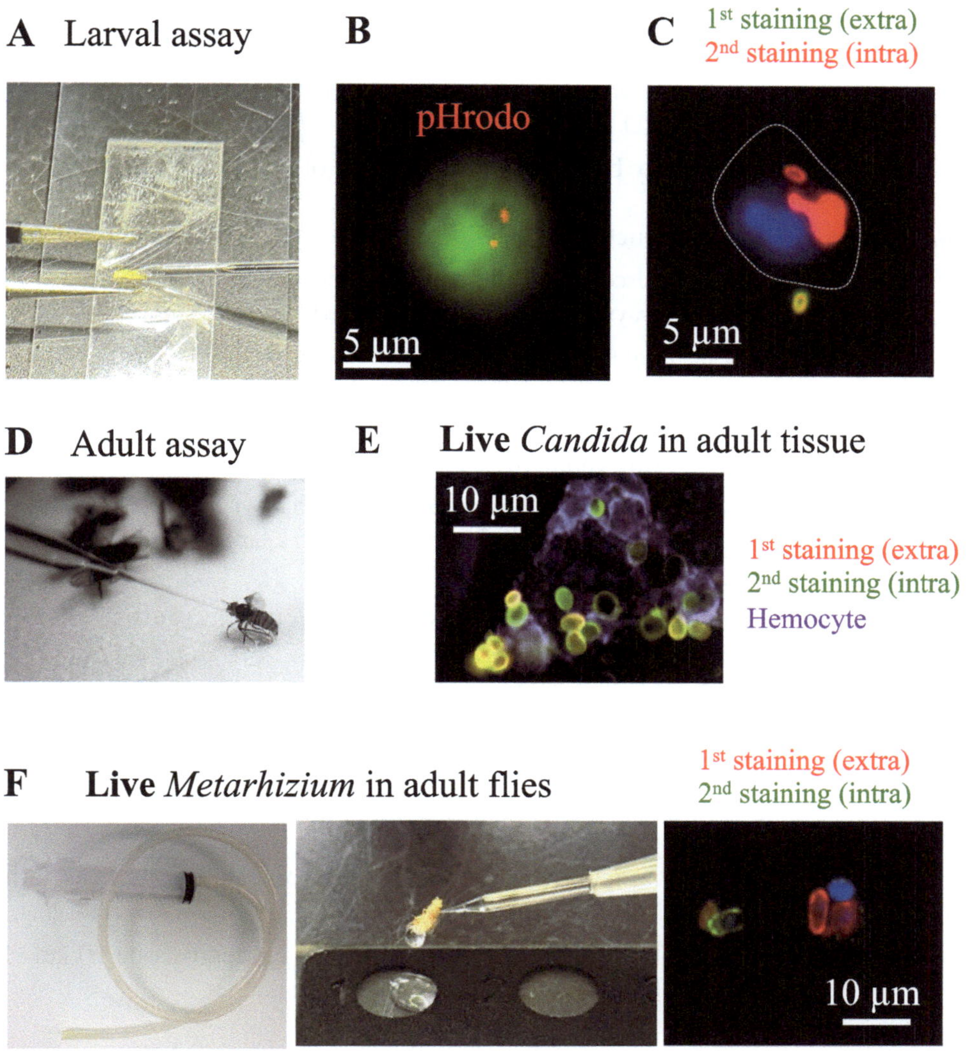

Fig. 2 Illustrations of in vivo phagocytosis assays in *D. melanogaster* larvae and adults. (**a–c**) In vivo phagocytosis assays in larvae. (**a**) Image illustrating the procedure of larval infection by injection, showing the position of the capillary (right) and the help of tweezers; the dried larva is stuck to double-sided tape on a slide. (**b**) Microscopy image of killed bacteria (labeled with red pHrodo) in a GFP-expressing plasmatocyte (from hmldelta-Gal4,UAS-GFP flies). (**c**) Microscopy image of bacteria with hemocytes, after in/out differential immunostaining: bacteria were first stained in green before permeabilization and then in red. Bacteria engulfed by hemocytes can only be stained in red, whereas extracellular bacteria are both red and green (yellow); the nucleus of an hemocyte is stained in blue (DAPI); the plasmatocyte membrane is outlined. (**d–f**) In vivo phagocytosis assays in adults. (**d**) Image illustrating the procedure of adult fly infection by injection. (**e**) Microscopy image of yeasts with hemocytes, after in/out differential immunostaining. Adult *Drosophila* tissues were dissected and fixed after *C. glabrata* injection. Hemocytes are shown in purple (P1 antibody). Yeasts cells were first stained in red before permeabilization and then in green. Ingested yeasts can only be stained in green, whereas extracellular yeasts are both green and red (yellow). (**f**) Left, adaptor for long capillaries with a syringe after a flexible tube, for adult hemolymph collection; middle, image illustrating the procedure of adult hemolymph collection by flooding the thorax with a large volume of PBS (about 20 µL), using the adaptor shown at the left image—hemocytes are collected in PBS on a 8-well microscope slide; right, *Metarhizium anisopliae* injection experiment, microscopy image of fungi with hemocytes, following in/out differential immunostaining procedure, fungi were first stained in red before permeabilization and then in green. Ingested spores can only be stained in green, whereas extracellular spores are both green and red (yellow); the nucleus of an hemocyte is stained in blue (DAPI)

4. *Metarhizium anisopliae* wild-type ARSEF 2575; cultured on PDA (Potato Dextrose Agar) plates at 25 °C for 7–14 days.
5. *Candida glabrata* wild-type ATCC 2001; overnight culture in YPD medium at 30 °C.
6. Any bacterial or fungal strain of interest.

2.2 Material for Microbial Solutions

1. GeneQuant pro Spectrophotometer (Amersham Biosciences).
2. Eresco 42 MF2, Cegelec (Source: Beryllium) generator of X-rays (for exposure to X-rays).
3. Precision wipes (Kimtech Science).

2.3 Material for Injection

1. Short capillaries: 3.5″ (Drummond).
2. Long capillaries: 7″ (Drummond).
3. Adaptor for long capillaries with a syringe after a flexible tube (in order to fill the capillary with a large volume, about 20 μL) (*see* Fig. 2f).
4. Flaming/Brown micropipette puller (Sutter Instrument, Model P-97).
5. Nanoject II Auto-Nanoliter Injector (Drummond).

2.4 Sample Preparation for Microscopy

1. Dumont #5 tweezers.
2. Diagnostic Microscope Slides Epoxy 8-Wells 6 mm Black (Thermo Scientific).
3. Plastic box with a layer of wet paper to use as a humid chamber.
4. Fluorescent Zeiss Axioscope 2 microscope (Carl Zeiss, Göttingen, Germany).

2.5 Antibodies

Primary antibodies detecting bacterial or fungal strains used for infection:

1. Rabbit antibodies against UV-killed wild-type *Pseudomonas aeruginosa* PA14 by Covalab; Dilution 1/200 [16] (*see* **Note 1**).
2. Antibodies raised against whole UV-killed *Candida* (from Karl Kuchler, Vienna) used against *C. glabrata* [8]; dilution 1/1000 (*see* **Note 1**).
3. Rabbit antibodies against UV-killed wild-type *M. anisopliae* ARSEF 2575 (From Wenhui Wang, SFHI, China); dilution 1/1000 (*see* Subheading 3.1).
4. Mouse monoclonal antibodies against the phagocytosis receptor Nimrod, expressed in hemocytes (P1, from István ANDÓ, Hungary) [22].
5. Matching secondary fluorescently labeled antibodies for detection.

6. FITC-labeled goat anti-rabbit secondary antibodies.
7. Cy3-labeled goat anti-rabbit secondary antibodies.

2.6 Media and Buffers

1. Lysogeny Broth LB-Miller: tryptone 10 g/L, yeast extract 5 g/L, NaCl 10 g/L in H_2O.
2. Bushnell Haas Broth (BHB): $MgSO_4$ 0.2 g/L, $CaCl_2$ 0.02 g/L, KH_2PO_4 1 g/L, K_2HPO_4 1 g/L, NH_4NO_3 1 g/L, $FeCl_3$ 0.05 g/L in H_2O.
3. Potato Dextrose Agar (PDA): dextrose 20 g/L, potato starch 4 g/L, Agar 15 g/L in H_2O.
4. Yeast extract peptone dextrose (YPD): peptone 20 g/L, dextrose 20 g/L, yeast extract 10 g/L in H_2O.
5. Phosphate-buffered saline (PBS): NaCl 8 g/L, KCl 0.2 g/L, Na_2HPO_4 1.44 g/L, KH_2PO_4 0.24 g/L in H_2O (pH = 7.4).

2.7 Reagents

1. Freund's adjuvant: 0.85 mL paraffin oil and 0.15 mL mannide monooleate (incomplete) ±1 mg of *Mycobacterium tuberculosis* (H37Ra, ATCC 25177), heat killed and dried (complete).
2. 0.1 M Na_2CO_3.
3. 16% paraformaldehyde aqueous solution.
4. Bovine serum albumin.
5. Tween 20 (Polysorbate 20).
6. Triton X-100.
7. Vectashield Antifade Mounting Medium with DAPI (Vector Laboratories).
8. Trypan blue 0.4% solution in PBS.
9. CML latex beads, 4% w/v, 0.3 μm (Thermo Fisher Scientific).
10. Fluorescein-5-isothiocyanate (FITC).
11. pHrodo™ Red succinimidyl ester (Invitrogen).
12. pHrodo™ Red *E. coli* BioParticles™ Conjugate for Phagocytosis (Invitrogen).

3 Methods

3.1 Development of Antibodies Against Metarhizium Anisopliae

1. *M. anisopliae* ARSEF 2575 is plated on PDA plate at 25 °C for 1 week.
2. 15 mL PBS with 0.01% Tween 20 are added to the plate, mixed, and a spore suspension is prepared by filtration through a precision wipe.
3. The suspension is transferred in a Petri dish on ice under a UV-C intensity of 1200 μW/cm² for 1 h (using a 15 W UV

lamp at a distance of 10 cm), allowing to kill microbes (*see* **Note 2**).

4. A small volume of UV-killed spore solution is plated to confirm the spores are dead.
5. Immunization was done by multiple subcutaneous injections (four times), each time with 250 million UV-killed spores, emulsified with complete Freund adjuvant at day 1, and emulsified with incomplete Freund adjuvant at day 14, 28, and 42, in the back of 4 New Zealand rabbits.
6. At day 49, animals were sacrificed, and their total blood was collected from heart. 1 mL of non-immunized rabbit serum was collected as a negative control.
7. Affinity purification of antisera has been done using Protein G chromatography, and 50 mg of antibodies were purified from each immunized rabbit.

3.2 Microbe Solution Preparation (Killed and Fluorescently Labeled Microbes, or Live Ones)

1. Killing can be achieved on fungal or bacterial cultures grown during the exponential phase (*see* **Note 2**):
   (a) By chemical fixation in PBS with 4% paraformaldehyde for 16 h at room temperature.
   (b) By heating the cultures at 65 °C for 60 min.
   (c) By administration of a UV-C light intensity of 1200 μW/cm^2 for 1 h (*see* Subheading 3.1, **step 3**).
   (d) By exposing cultures to X-rays at 2 krad/min (20 Gy/min) for 6 h corresponding to a total exposure dose of 720 krad.

   Killed cells are washed twice in PBS and plated to ensure nonviability, on LB, BHB, PDA, or YPD agar plates, according to the microbial species (*see* Subheading 2.1).
2. Staining of killed microbes can be done using FITC or pHrodo. The principle of pHrodo is that this dye is nonfluorescent at neutral pH and exhibits increasing fluorescence (red or green) as the pH becomes more acidic, such as that encountered in the mature phagolysosome. Thus, there is no need for quenching, and only internalized bacteria that are being located in the final degradation compartment are visualized (*see* **Note 3**). FITC labeling is done by washing microbes (10^8 cells/mL) with a 0.1 M sodium carbonate (Na_2CO_3) solution at pH = 9.2 (pH is important) with 0.05% Tween 20 followed by incubation in the same solution with 100 μg/mL FITC, overnight at 4 °C or 2 h at room temperature, in the dark (from a stock solution of 2 mg/mL FITC in DMSO). Wash extensively with PBS + 0.01% Tween 20 (four times with an equal volume). Check the final cell suspension microbe concentration by direct counting using a hemocytometer.

Microbes can be labeled with pHrodo, following the instructions from Invitrogen. Solutions at 5 × 10^{10} bacteria/mL in PBS (dose used for *Pseudomonas aeruginosa* PA14) are aliquoted and kept frozen at −20 °C before use (*see* **Note 4**).

3. Live microbes: Fresh overnight cultures are used for live assays. Optical density (OD) at 600 nm was adjusted to a value of 20 in PBS for the bacteria *P. aeruginosa* or *S. marcescens*; for fungi, 4000 *Candida glabrata* yeasts in a 0.01% Tween 20 solution in PBS (*see* **Note 5**) are injected whereas 3500 *M. anisopliae* spores are injected (69 nL of a 5 × 10^7 spores/mL solution).

3.3 Animal Bleeding and Dissection for Visualization of Internalized Bacteria on Tissue-Resident Plasmatocytes (See Note 6)

1. Larvae are bled by dissecting it with tweezers and shaking it on the solution into the wells, before discarding the cadaver: hundreds of hemocytes will be found in each well (*see* **Note 7**).
2. Adult hemolymph collection needs more preparation. First, cut a small part of the abdomen, using tweezers (a hole is enough). Second, take the adaptor described in Subheading 2.3, **step 3** and fill a long capillary with 20 μL PBS that can be injected in the thorax: this PBS will go out from the opened abdomen with hemocytes. Take this solution into a well from an 8-well microscope slide that was previously filled with 20 μL of PBS (*see* Fig. 2f).
3. Adult tissue dissection consists in removing the head and the thorax, keeping the fly abdomen, dissecting the abdomen from the ventral side, and then removing all internal tissues (reproductive and digestive tract). Most hemocytes will still be present after dissection, lining the respiratory epithelia of the head and thorax [23].

3.4 Ex Vivo Phagocytosis Assays (See Note 8)

1. Place a volume of 20 μL of a microbial solution (either killed or live microbes) in each well of an 8-well microscope slide.
2. Bleed one animal (a third instar wandering larva or an adult fly) per well (*see* **Note 9**).
3. Incubate for 40 min in a humid chamber (*see* **Note 10**).
4. For FITC-labeled particles, mount the slides on a 0.4% trypan blue solution in PBS (*see* **Note 11**), and incubate for 10–30 min before observation by microscopy (*see* Subheading 3.8, **step 3**).
5. For pHrodo-labeled particles, wash for 10 min in 20 μL of a 0.5 μg/mL DAPI solution in PBS, mount the slides on PBS, and analyze by live microscopy (*see* Subheading 3.8, **step 3**). For non-labeled microbes, follow the steps described for sample preparation for microscopy (*see* Subheading 3.6).

3.5 In Vivo Phagocytosis Assays

1. Inject third instar wandering larvae (*see* **Note 12**) or adult flies with 27.6 nL of a microbial solution (either killed or live microbes), using a Nanoject apparatus (Drummond).
2. Incubate for 60 min (*see* **Note 13**).
3. Bleed one larva (*see* **Notes 9** and **14**) or dissect tissue from one fly per well, into 20 μL of PBS on an 8-well microscope slide.
4. Incubate for 25 min in a humid chamber (*see* **Note 10**).
5. For pHrodo-labeled particles, wash for 10 min in 20 μL of a 0.5 μg/mL DAPI solution in PBS, mount the slides on PBS, and analyze by live microscopy (*see* Subheading 3.8, **step 3**). For non-labeled microbes, follow the steps described on Subheading 3.8.

3.6 Opsonization Assay on Live Microbes (Fig. 1b)

The opsonization assay involves a complex procedure. First, the live microbes are incubated in vitro with cell-free hemolymph collected from larvae of either wild-type or mutant for gene encoding a potential opsonin. The microorganisms will become coated with the opsonins if present in the collected hemolymph. These potentially opsonized bacteria are then injected in opsonin-deficient larvae, and the phagocytic index is measured on bled plasmatocytes after an incubation period. If the mutant gene product is really required for opsonization, then the phagocytic index will be lower when the larvae have been injected with the microorganisms incubated first with the opsonin-deficient hemolymph than when injecting the opsonized microbes incubated at the first step of the procedure with wild-type hemolymph [16].

1. Prepare 150 μL of a microbial solution in a 1.5 mL microtube (*see* **Note 3**).
2. Bleed 20 third instar larvae directly into the bacterial solution (*see* **Note 15**).
3. Incubate at room temperature for 30–45 min (to allow potential opsonization to take place).
4. Centrifuge at 500 rcf for 15 min and transfer the supernatant in a new tube (*see* **Note 16**).
5. Centrifuge at 3500 rcf for 15 min to retrieve microbes in the pellet (*see* **Note 17**).
6. Resuspend the pellet in 10 μL PBS.
7. Inject third instar larvae (*see* **Note 12**) with 32.2 nL of the last solution obtained, using a Nanoject apparatus: the recipient larvae used are mutant for the putative opsonin gene.
8. Incubate for 60 min (*see* **Note 13**).
9. Bleed one larva per well, into 20 μL of PBS on an 8-well microscope slide.

10. Incubate for 25 min in a humid chamber (*see* **Note 10**).
11. Follow the steps described on Subheadings 3.7 and 3.8.

3.7 In/Out Differential Immunostaining Procedure for Unlabeled Microbes

1. Fix hemocytes (from bled larva or adults) in 1% paraformaldehyde in PBS for 10 min, and fix dissected adult tissues in 4% paraformaldehyde in PBS for 30 min, in a humid chamber (*see* **Note 18**).
2. Wash twice in PBS for 5 min.
3. Block for 30 min in a PBS solution with 2% Bovine Serum Albumin (BSA).
4. Stain the extracellular antigen with a primary antibody solution (against the microbe of interest), in PBS with 2% BSA overnight at 4 °C. This antibody will not reach the intracellular antigen because the cells are not permeabilized at this step.
5. Wash twice in PBS for 5 min.
6. Incubate with a FITC-labeled secondary antibody in a PBS solution with 2% BSA for 2 h at room temperature.
7. Wash twice in PBS for 5 min.
8. Permeabilize for 30 min in a PBS solution with 0.1% Triton X-100 and 2% BSA (*see* **Note 19**).
9. Stain the intracellular antigen with a solution containing the same primary antibody as used at **step 4**, in PBS with 0.1% Triton X-100 and 2% BSA overnight at 4 °C (*see* **Note 20**).
10. Wash twice in PBS for 5 min.
11. Incubate with a Cy3-labeled secondary antibody in a PBS solution with 0.1% Triton X-100 and 2% BSA for 2 h at room temperature.
12. Wash twice in PBS for 5 min.
13. Mount the slide in Vectashield with or without DAPI.

3.8 Sample Analysis by Microscopy

1. The samples are analyzed using a Zeiss Axioscope 2 fluorescent microscope or equivalent. In the case of red-fluorescent pHrodo-labeled particles, fluorescent microbes are considered as intracellularly located (*see* **Note 21**).
2. After in/out differential immunostaining, red and non-green particles are considered as intracellularly localized (*see* **Note 22** and Fig. 2). The number of red fluorescent bacteria that are not green fluorescent is counted in each DAPI-positive hemocyte.
3. Between 50 and 100 cells are analyzed per animal: the number of intracellular microbes is counted for each hemocyte (DAPI-positive or P1-positive, depending on the protocol), and the phagocytic index is calculated (% of phagocytes containing at least 1 bacterium) × (mean number of bacteria per positive cell).

After several independent experiments, each using at least 6 animals per condition, the non-parametric Mann-Whitney test is used for statistical analysis.

4 Notes

1. Antibodies against microbes.

   The antibodies against microbes can work only if they can recognize surface epitopes: indeed, they should bind their epitopes even in the absence of permeabilization, a condition required for the in/out differential immunostaining procedure.

2. Killed versus live microbes.

   Killing procedure can affect microbial surface. UV and X-ray killing better preserve the surface properties of microbes than chemical fixation and heating [24, 25]. Killed microbes may be more efficiently engulfed as compared to live ones, because some microbes have established active mechanisms to escape detection or engulfment by phagocytes: some of these mechanisms may be affected when microbes are killed. It is what we clearly found for *P. aeruginosa* [5, 16].

3. The pHrodo labeling needs to use live plasmatocytes.

   The pHrodo technique is performed using live plasmatocytes as fixation leads to the loss of the fluorescent signal. Thus, it is fast but requires immediate analysis of the samples. If one fails to observe any red signal in a mutant background, it does not imply that internalization is blocked, as only the maturation step might be affected. In such cases, it is therefore valuable to use the other techniques that allow monitoring internalization such as FITC-labeled microorganisms and quenching of the extracellular bacteria or the in/out differential staining procedure.

4. Dose of microbes to use in phagocytosis assays.

   Depending on the phagocytosis efficiency of a given microbe, it is necessary to determine the optimal dose to use. For instance, we observed that *E. coli* and *S. aureus* are much more efficiently phagocytosed as compared to *S. marcescens* and *P. aeruginosa*.

5. Addition of Tween 20 in fungal solutions and sedimentation.

   A low concentration of Tween 20 helps preventing fungal particle aggregation. A PBS solution without Tween 20 is used in the case of bacterial solutions. Before injection, fungal solutions need to be mixed because of the fast sedimentation of fungi. This is less critical for bacterial solutions that settle down much slower.

6. Retrieved hemocytes from animal bleeding are mainly the circulating ones.

Among the two populations of larval hemocytes, the circulating ones are mainly retrieved by bleeding, and the sessile ones are mainly lost, adherent to the larval tissues. Since adult hemocytes are mainly sessile, a similar population of cells is likely retrieved by bleeding or observed after adult tissue dissection.

7. Larvae need to be synchronized.

   It is critical to retrieve hemocytes from the third larval stage just after the peak of ecdysone expression that occurs at the mid-third instar and changes the behavior of larvae. The larvae have to be taken when they are out of the food (wandering stage) and before they stop to move. Even if the larvae are well synchronized, some can develop slower, especially if their density is too high. In fact, exposure to ecdysone is required for the hemocyte phagocytic activity and can thus modify the level of phagocytosis [26].

8. Ex vivo versus in vivo phagocytosis assay.

   This ex vivo procedure can be less accurate as compared to in vivo phagocytosis assay for two reasons:

   (a) The composition of the hemolymph is different than PBS, and the buffer used may change the cell physiology, blocking some biological processes like engulfment. However, in our experience, ex vivo phagocytosis assays in PBS or Schneider's *Drosophila* medium yield similar results.

   (b) Other organs (like the fat body) may secrete some factors that may influence the phagocytosis efficiency (for instance, opsonins) that may be required for the detection of some microbes by phagocytes.

9. Larval versus adult hemocytes.

   In adult flies, there are much fewer hemocytes retrieved (tens per fly) as compared to larvae (hundreds per larva), and the hemocytes are more autofluorescent. In addition, it is more time-consuming to retrieve adult hemolymph than the larval one (*see* Fig. 2f). However, in some cases, it may be more relevant to study phagocytosis by adult hemocytes, as one cannot formally exclude that the properties of hemocytes differ at both developmental stages. Indeed, most adult hemocytes are sessile: their attachment to tissues may modify the efficiency with which they are able to interact with microbes. Of note, we never tested adult hemocytes on ex vivo phagocytosis assays, only on in vivo assays, but it should be feasible.

10. Incubation time to allow the hemocytes to settle down.

    A humid chamber is important to avoid desiccation of the samples. During this incubation time, hemocytes will settle down. In the case of ex vivo assays, they will engulf microbes at the same time. In the case of in vivo assays on adult dissected tissues, this incubation time is not needed.

11. Trypan blue can quench the FITC fluorescence from extracellular particles.

    Trypan blue is needed to quench the FITC fluorescence of extracellular particles. Thus, only the intracellular particles are able to emit fluorescence. This protocol allows quantifying the engulfment of particles.

12. Larval injection.

    Larval injection is not easy. To avoid killing the larvae, use a capillary as thin as possible. In addition, the larvae can be easily stuck on a dry surface (e.g., on a glass microscopy slide) after making the larva itself dryer (by absorbing liquid from its surface on a paper). One may use double-sided tape stuck on a slide. Then, the most convenient way to inject is to keep a position for the capillary almost parallel to the larval antero-posterior axis and to inject in the middle of the body. Tweezers can be helpful to keep the larva immobilized (*see* Fig. 2a).

13. Incubation time for in vivo phagocytosis.

    During this 1 h incubation step, the hemocytes in injected larvae or adult flies will engulf microbes.

14. Monitoring phagocytosis in situ.

    To monitor in vivo phagocytosis, one possibility is to bleed the injected insect using the same techniques as for the ex vivo assays. The other option is to dissect the tissues to visualize the hemocytes in situ, a strategy that requires the observation of multiple adult flies to obtain high enough numbers for meaningful statistical analysis.

15. Type of samples to use in opsonization assays.

    The different samples to compare in an opsonization assay may be WT larvae to larvae mutant for a gene encoding a putative opsonin, for instance, Tep4 [16]. As control conditions, one can use flies deficient for phagocytosis, either by "latex" bead injection in WT flies or by using mutant flies impaired for phagocytosis [4, 14, 15].

16. Low-speed centrifugation.

    Bacteria bound by opsonins are retrieved in the supernatant. The low-speed of centrifugation is essential to ensure that bacteria are not also pelleted with cells and tissues. This step was done successfully with bacteria but will need to be optimized for fungal spores or yeasts, which are bigger and sediment faster.

17. High-speed centrifugation.

    At this step, microbes bound to potential opsonins are retrieved in the pellet.

18. Immunofluorescence troubleshooting with permeabilization.

    One may observe some intracellular staining even when cells are not permeabilized. This is because formaldehyde fixation can affect membrane integrity [27]. To avoid this, it might be better to use a 1% diluted formaldehyde solution and to shorten the period of fixation.

19. Permeabilization of cell membranes.

    Triton X-100 and NP-40 are harsh detergents, as compared as Tween 20, saponin, or digitonin that are mild ones. Although an efficient permeabilization can significantly improve antibody access to antigens in the cytoplasm, a harsh detergent may disrupt some antigens. Triton X-100 was suitable for all antigens described here.

20. Immunostaining with additional antibodies.

    After the permeabilization step following the first antibody staining, if needed, it is possible to simultaneously use an additional primary antibody raised in another animal species, different from the anti-microbe antibody. Then, a secondary antibody labeled with another fluorescent dye can be used on the next step. If the antibody was raised in the same animal from the anti-microbe antibody, for instance, in the case of the plasmatocyte-specific mouse P1 antibody, the whole procedure can be done in parallel, to get yellow plasmatocytes and yellow extracellular microbes (*see* Fig. 2e). The fact that we cannot discriminate the color is not a problem to count engulfed particles.

21. Fluorescence detection on pHrodo-labeled microbes.

    pHrodo exhibits increasing fluorescence (red or green) as the pH becomes more acidic, such as that encountered in the mature phagolysosome. Thus, every fluorescent microbe can be considered as internalized by hemocytes into matured phagolysosome. Of note, as contrary as other protocols described here, the usage of pHrodo will not help to quantify every engulfed particle, but only the ones present in maturing phagolysosomes. In practical, there is always a weak background fluorescence, but its intensity is much lower than the signal from engulfed microbes.

22. Fluorescence detection of microbes after in/out differential immunostaining.

    The detected fluorescence of extracellular microbes after permeabilization is often only due to the first immunostaining step. This may be due to the saturation of epitopes during the first immunostaining step (before permeabilization), competing for a possible staining on the second immunostaining step (*see* Fig. 2f). Of note, the secondary antibodies may be inverted between the first step and the second step of immunostaining (compare Fig. 2c–f).

Acknowledgments

We would like to thank François Lapraz (Institute of Biology Valrose, Nice, France) and Gábor Csordás (Institute for Genetics, University of Cologne) for their advice for adult hemolymph collection and Marion Draheim for critical reading of the manuscript. DF's teamwork at SFHI is partially funded through the 1,000 Talent Program of Global experts of China.

References

1. Lemaitre B, Hoffmann J (2007) The host defense of Drosophila melanogaster. Annu Rev Immunol 25:697–743
2. Gold KS, Brückner K (2015) Macrophages and cellular immunity in Drosophila melanogaster. Semin Immunol 27(6):357–368
3. Lanot R, Zachary D, Holder F, Meister M (2001) Postembryonic hematopoiesis in drosophila. Dev Biol 230:243–257
4. Kocks C, Cho JH, Nehme N, Ulvila J, Pearson AM, Meister M et al (2005) Eater, a transmembrane protein mediating phagocytosis of bacterial pathogens in drosophila. Cell 123:335–346
5. Nehme NT, Liegeois S, Kele B, Giammarinaro P, Pradel E, Hoffmann JA et al (2007) A Model of bacterial intestinal infections in *Drosophila melanogaster*. PLoS Pathog 3:e173
6. Limmer S, Haller S, Drenkard E, Lee J, Yu S, Kocks C et al (2011) *Pseudomonas aeruginosa* RhlR is required to neutralize the cellular immune response in a *Drosophila melanogaster* oral infection model. Proc Natl Acad Sci U S A 108:17378–17383
7. Nehme NT, Quintin J, Cho JH, Lee J, Lafarge MC, Kocks C et al (2011) Relative roles of the cellular and humoral responses in the Drosophila host defense against three gram-positive bacterial infections. PLoS One 6:e14743
8. Quintin J, Asmar J, Matskevich AA, Lafarge MC, Ferrandon D (2013) The Drosophila toll pathway controls but does not clear *Candida glabrata* infections. J Immunol 190:2818–2827
9. Pearson AM, Baksa K, Rämet M, Protas M, McKee M, Brown D et al (2003) Identification of cytoskeletal regulatory proteins required for efficient phagocytosis in Drosophila. Microbes Infect 5(10):815–824
10. Elrod-Erickson M, Mishra S, Schneider D (2000) Interactions between the cellular and humoral immune responses in *Drosophila*. Curr Biol 10:781–784
11. Nazario-Toole AE, Wu LP (2019) Assessing the cellular immune response of the fruit Fly, *Drosophila melanogaster*, using an in vivo phagocytosis assay. J Vis Exp 146
12. Hoffmann D (1976) Role of phagocytosis and soluble antibacterial factors in experimental immunization of Locusta migratoria. C R Acad Hebd Seances Acad Sci D 282:1021–1024
13. Rutschmann S, Kilinc A, Ferrandon D (2002) The *Toll* pathway is required for resistance to gram-positive bacterial infections in *Drosophila*. J Immunol 168:1542–1546
14. Defaye A, Evans I, Crozatier M, Wood W, Lemaitre B, Leulier F (2009) Genetic ablation of drosophila phagocytes reveals their contribution to both development and resistance to bacterial infections. J Innate Immun 1:322–334
15. Charroux B, Royet J (2009) Elimination of plasmatocytes by targeted apoptosis reveals their role in multiple aspects of the drosophila immune response. Proc Natl Acad Sci U S A 106:9797–9802
16. Haller S, Franchet A, Hakkim A, Chen J, Drenkard E, Yu S et al (2018) Quorum-sensing regulator RhlR but not its autoinducer RhlI enables pseudomonas to evade opsonization. EMBO Rep 19(5):e44880
17. Avet-Rochex A, Bergeret E, Attree I, Meister M, Fauvarque MO (2005) Suppression of drosophila cellular immunity by directed expression of the ExoS toxin GAP domain of *Pseudomonas aeruginosa*. Cell Microbiol 7:799–810
18. Avet-Rochex A, Perrin J, Bergeret E, Fauvarque MO (2007) Rac2 is a major actor of drosophila resistance to *Pseudomonas aeruginosa* acting in phagocytic cells. Genes Cells 12:1193–1204
19. Sinenko SA, Mathey-Prevot B (2004) Increased expression of drosophila tetraspanin, Tsp68C, suppresses the abnormal proliferation of ytr-deficient and Ras/Raf-activated hemocytes. Oncogene 23:9120–9128

20. Haller S, Limmer S, Ferrandon D (2014) Assessing pseudomonas virulence with a non-mammalian host: *Drosophila melanogaster*. Methods Mol Biol 1149:723–740
21. Lestradet M, Lee K-Z, Ferrandon D (2014) Drosophila as a model for intestinal infections. Methods Mol Biol 1197:11–40
22. Kurucz E, Markus R, Zsamboki J, Folkl-Medzihradszky K, Darula Z, Vilmos P et al (2007) Nimrod, a putative phagocytosis receptor with EGF repeats in drosophila plasmatocytes. Curr Biol 17:649–654
23. Bosch PS, Makhijani K, Herboso L, Gold KS, Baginsky R, Woodcock KJ, et al (2019) Blood cells of adult Drosophila do not expand, but control survival after bacterial infection by induction of Drosocin around their reservoir at the respiratory epithelia. bioRxiv. 578864
24. Datta SK, Okamoto S, Hayashi T, Shin SS, Mihajlov I, Fermin A et al (2006) Vaccination with irradiated listeria induces protective T cell immunity. Immunity 25:143–152
25. Wheeler RT, Fink GRA (2006) Drug-sensitive genetic network masks fungi from the immune system. PLoS Pathog 2(4):e35
26. Regan JC, Brandão AS, Leitão AB, Mantas Dias AR, Sucena E, Jacinto A et al (2013) Steroid hormone signaling is essential to regulate innate immune cells and fight bacterial infection in drosophila. PLoS Pathog 9(10):e1003720
27. Pelts M, Pandya SM, Oh CJ, Model MA (2011) Thickness profiling of formaldehyde-fixed cells by transmission-through-dye microscopy. BioTechniques 50(6):389–396

Chapter 6

Analysis of Cellular Immune Responses in Lepidopteran Larvae

Andrea Becchimanzi, Ilaria Di Lelio, Francesco Pennacchio, and Silvia Caccia

Abstract

The immune system in insects is innate and relies both on humoral components and cellular response, which includes phagocytosis, nodulation, and encapsulation. Humoral and cellular defenses are inextricably linked in a complex immune reaction, modulated by a number of cross-talking pathways, which can result in a remarkable variability of any measurement. Protocols to assess the immune response are quite critical and should be carefully standardized, in order to achieve a desirable level of reproducibility. Here we report methods for measuring the cellular immune response in insects, using lepidopteran larvae as model. These protocols have been developed starting from the fragmented information available in the literature, which has been integrated and refined on the basis of our experience on the study of insect immunosuppression by parasitoids and pathogens.

Key words Insect immunity, Hemocytes, Phagocytosis, Nodulation, Encapsulation

1 Introduction

Insects lack an adaptive immune response like that of vertebrates but show immune priming [1, 2] and a wealth of innate immune defenses, such as physical barriers (e.g., integument and intestine) preventing the entrance into the body cavity (hemocoel) of foreign invaders, and an array of both humoral and cellular reactions, activated by wounding, tissue damage, and pathogens or parasites [3–6].

Humoral defenses include the synthesis of antimicrobial peptides [7, 8], reactive oxygen and/or nitrogen species [9, 10] and the activation of complex enzymatic cascades that regulate coagulation and melanization of hemolymph [11–13]. On the other hand, cellular defense is mediated by circulating immune cells (i.e., hemocytes) and includes phagocytosis, nodulation, and encapsulation [14–21]. Specific immune responses are triggered

Andrea Becchimanzi and Ilaria Di Lelio contributed equally to this work.

Federica Sandrelli and Gianluca Tettamanti (eds.), *Immunity in Insects*, Springer Protocols Handbooks,
https://doi.org/10.1007/978-1-0716-0259-1_6,

by the recognition of non-self organisms and objects or altered self, a critical event that is mediated by an array of humoral factors circulating in the hemolymph and/or cellular receptors associated with hemocytes [3, 22]. The invaders are detected by the recognition of pathogen-associated molecular patterns (PAMPs) that are present on their surface (e.g., lipopolysaccharides in Gram-negative bacteria, peptidoglycan and lipoteichoic acid in Gram-positive bacteria, and β-1,3 glucans in fungi) [4, 20]. In Lepidoptera, a number of so-called pattern recognition receptors (PRRs) are present on the surface of both hemocytes and fat body cells or circulate in the hemolymph (humoral PRRs) [4, 20]. In the case of hemocyte-associated PRRs, their binding to PAMPs activates specific intracellular signaling pathways controlling the cellular response, which, in some cases, can be also mediated by humoral PRRs, such as opsonizing factors [4, 20].

1.1 Phagocytosis

Phagocytosis is the process by which an individual immune cell recognizes, binds, internalizes, and destroys relatively small non-self particles and microorganisms (usually larger than 0.5 μm in diameter up to several μm, depending on non-self shape and hemocyte size) [23–27]. Phagocytosis is rapid and, unlike nodulation or encapsulation, is a renewable response whereby hemocytes repeatedly internalize and degrade pathogens [28]. In insects, the hemocyte's capacity to phagocyte various bacteria differs. It has been reported that the Gram-negative bacteria *Escherichia coli* are more readily phagocytized than the Gram-positive bacteria *Staphylococcus aureus* by immune cells in different insect species [29–33].

The class of hemocytes reported to be phagocytic vary among different insect orders [3, 29, 32, 34–37]. In Lepidoptera, granulocytes and plasmatocytes are the only hemocyte types reported to be phagocytic [3, 38], while in *Drosophila melanogaster* plasmatocytes are the main phagocytic immune cells [34].

1.2 Nodulation

Nodulation is an insect cellular defense reaction triggered by severe bacterial infections (e.g., in the larvae of the lepidopteran *Galleria mellonella*, it is triggered when the number of bacteria for μL of hemolymph exceeds 10^3), or when invaders cannot be phagocytized because of their size [39–41]. This cellular response consists in the formation of multicellular hemocyte aggregates that quickly entrap and clear large clumps of invading bacteria [14, 15, 40, 41]. The nodulation process is usually completed by the activation of the phenoloxidase cascade that leads to melanization of microorganisms by mature nodules [40, 42].

1.3 Encapsulation

The hemocoel invasion by larger intruders (e.g., protozoans, metazoan parasites, nematodes, and parasitoid eggs or larvae) triggers within minutes a cooperative reaction by hemocytes, which bind to the foreign target and form multiple cell layers, encapsulating the invader, which is eventually suppressed by melanization [41, 43], by local production of cytotoxic-free oxygen and nitrogen, or by asphyxiation [44–48]. In *Drosophila*, lamellocytes (large flat hemocytes present in *Drosophila* hemolymph) appear to be the predominant cells involved in capsule formation [10, 15]. In Lepidoptera, encapsulation starts with granulocytes, which form the first layer in contact with the non-self object, followed by the apposition of multiple layers of plasmatocytes and by a single layer of granulocytes that terminates the process [49].

Encapsulation and nodulation responses are based on a cooperative behavior of hemocytes and share both functional and molecular features [3, 40, 41, 50]. Indeed, both processes involve the formation of multilayered hemocyte aggregates around the non-self intruders, which are suppressed by deposition of toxic metabolites (e.g., melanin).

Hemocytes are classified into distinct types based on morphological and functional characteristics and have different names in different insect orders [3, 29, 51–53]. Four hemocyte types have been described in lepidopteran larvae based on morphology and functions [3, 54, 55]: plasmatocytes and granulocytes are involved in phagocytosis, nodulation, and capsule formation; oenocytoids produce enzymes (e.g., phenoloxidase) involved in the melanization cascade and spherule cells whose immune function remains unclear [3, 55].

Here we present simple laboratory protocols for the quantitative assessment of the cellular immune responses in lepidopteran larvae, that are based on the fragmented information available in the literature and refined in our laboratory, which are currently used to assess the hemocyte behavior in caterpillars as affected by parasitoids and pathogens.

2 Materials

Prepare solutions with ultrapure water and analytical grade reagents. Prepare and store reagents at room temperature unless otherwise indicated.

2.1 *In Vitro* Phagocytosis Assay

1. Forceps (straight shape, tip dimensions: 0.05 × 0.02 mm, length: 11 cm) and microscissors (curved blades, extra fine points, length: 100 mm).
2. Parafilm.
3. Hemocytometer.
4. Phosphate-buffered saline (PBS) (137 mM NaCl, 2.7 mM KCl, 10 mM phosphate buffer, pH 7.4). To prepare 1 L: dissolve 8 g of NaCl (molecular weight (MW) 58.44 g/mol), 0.20 g of KCl (MW 74.56 g/mol), 3.58 g of $Na_2HPO_4{\cdot}12H_2O$

(MW 358.14 g/mol), and 0.27 g of KH_2PO_4 (MW 136.09 g/mol) in 900 mL of water. Adjust pH to 7.4 with 1 M HCl, and bring to 1 L with distilled water. Autoclave the solution.

5. Suspensions of 2×10^7 of fluorescein conjugated *E. coli* (Gram-negative) cells (K-12 strain BioParticles®, fluorescein conjugate, Invitrogen), and fluorescein conjugated *S. aureus* (Gram-positive) cells (Wood strain, BioParticles® fluorescein conjugate, Invitrogen) per mL of PBS. Fluorescent bacteria are provided as 10 mg of lyophilized powder, which contains approximately 3×10^8 *E. coli* or *S. aureus* cells per mg. Reconstitute and store as described by manufacturers. Briefly, add 10 mL of PBS to prepare a stock of 3×10^8 cells/mL and vortex vigorously the particles (3×15 s at the highest setting). Reconstituted suspensions can be stored at 4 °C for several weeks, with the addition of sodium azide to a final concentration of 2 mM. Working dilutions (2×10^7 cells/mL) can be prepared just before use by proper dilution of the stock with PBS. Protect suspensions of fluorescent BioParticles® conjugates from light by wrapping the tubes with aluminum foil.
6. Fluorescence microscope (10× eyepieces and 40× objective for a 400× overall magnification) equipped with FITC (fluorescein isothiocyanate) filter (excitation wavelength: 475 nm, emission wavelength: 530 nm).
7. Manual cell counter.

2.2 In Vivo Nodulation Assay

1. Forceps (straight shape, tip dimensions: 0.05×0.02 mm, length: 11 cm) and microscissors (curved blades, extra fine points, length: 100 mm).
2. Parafilm and autoclaved filter paper.
3. Cylindrical plastic vials measuring approximately 30×50 mm and cotton wool.
4. Hemocytometer.
5. Hamilton microsyringe 1701 RN SYR (10 μL, 26s gauge, 55 mm long, point style 2).
6. LB-agar plate with *E. coli* colonies.
7. Sterile loops.
8. Spectrophotometer (equipped for 600 nm wavelength measurements).
9. 70% solution of ethanol in water.
10. Autoclaved water.
11. PBS (*see* **item 4** in Subheading 2.1).
12. MEAD anticoagulant buffer (98 mM NaOH, 145 mM NaCl, 17 mM EDTA, 41 mM citric acid, pH 4.5). To prepare 50 mL:

add 0.196 g of NaOH (MW 39,997 g/mol) and 0.424 g of NaCl (MW 58.44 g/mol) in 2.05 mL of 1 M citric acid (MW 192,124 g/mol) and 1.7 mL of a 0.5 M water solution of EDTA (adjusted at pH 8 with NaOH). Add distilled water to 40 mL and adjust pH to 4.5 with 1 M HCl. Then adjust the volume to 50 mL and autoclave the solution.

13. Suspension of 2 × 10^6 *E. coli* cells/mL of PBS. Prepare LB (Luria-Bertani) liquid medium adding 10 g of tryptone, 10 g of NaCl and 5 g of yeast extract in 950 mL of water. Shake until the solutes are dissolved. Adjust the pH to 7.0 with 5 M NaOH and adjust to a final volume of 1 L with water. Autoclave the medium. Inoculate 5 mL of LB broth with several *E. coli* colonies (picked with a sterile loop from the LB-agar plate) into a 15 mL plastic tubes. Incubate overnight at 37 °C, under shaking. Measure the optical density at 600 nm (OD600) with a spectrophotometer and dilute bacteria to OD600 = 0.5 (equivalent to 4 × 10^8 cells/mL) with LB. Spin the bacterial suspension in 15 mL plastic tubes at 4000 × *g* for 10 min at room temperature. Eliminate the supernatant and resuspend bacteria in 10 mL of PBS to obtain a 2 × 10^8 cells/mL suspension. Dilute the suspension 1:100 (v/v) with PBS to obtain the working solution of 2 × 10^6 cells/mL.
14. Hemocytometer.
15. Light microscope (10× eyepieces and 40× objective for a 400× overall magnification).
16. Manual cell counter.

2.3 In Vivo Encapsulation of Chromatography Beads

1. Forceps (straight shape, tip dimensions: 0.05 × 0.02 mm, length: 11 cm) and microscissors (curved blades, extra fine points, length: 100 mm).
2. Cylindrical plastic vials measuring approximately 30 × 50 mm and cotton wool.
3. 70% solution of ethanol in water.
4. Sterile water.
5. PBS (*see* **item 4** in Subheading 2.1).
6. MEAD anticoagulant buffer (*see* **item 12** in Subheading 2.2).
7. Hamilton microsyringe 702 RNR SYR (25 μL, gauge 22s, length 55 mm, needle 2).
8. Parafilm.
9. Disposable Petri dishes (diameter: 10 cm).
10. CM Sepharose® Fast Flow chromatography beads (Sigma Aldrich) (*see* **Note 1**).
11. Cell strainers with nylon mesh (70 μm) fitting standard 50 mL centrifuge tubes.

12. 50 mL centrifuge tubes.
13. Stereomicroscope (10× eyepieces and 1–8× objective).
14. Inverted microscope (10× eyepieces and 40× objective for a 400× overall magnification).
15. 96-well cell culture plates, flat bottom.
16. Manual cell counter.

2.4 *In Vivo* Encapsulation of a Nylon Thread

1. Forceps (straight shape, tip dimensions: 0.05 × 0.02 mm, length: 11 cm), scalpel, and fine needles (>28 gauge).
2. 70% solution of ethanol in water.
3. Nylon fishing line (0.08–0.2 mm of diameter).
4. Fine sand paper (220 grit).
5. Horizontal laminar flow hood equipped with a UV lamp.
6. Cylindrical plastic vials measuring approximately 30 × 50 mm and cotton wool.
7. Stereomicroscope (10× eyepieces and 1–8× objective) mounting a digital camera.
8. Image analysis software (e.g., GIMP, ImageJ, Photoshop, etc.).

3 Methods

The assays described below have been optimized for fifth and sixth larval instars of *Spodoptera littoralis* (Lepidoptera, Noctuidae), reared on artificial diet as elsewhere described [56]. In Subheading 4, we provide essential information for adapting the protocols when using other species.

3.1 *In Vitro* Phagocytosis of Bacteria

1. Anaesthetize *S. littoralis* larvae using carbon dioxide (CO_2) for 2–3 min or by chilling on ice for 30 min.
2. Place a larva on a parafilm piece, with the legs facing upward.
3. Cut a thoracic leg, and gently squeeze the larval body to collect 10 μL of exuding hemolymph with a micropipette, and immediately transfer hemolymph into a microcentrifuge tube (1.5 mL) containing 10 μL of ice-cold sterile PBS.
4. Add 4 μL of 2×10^7/mL fluorescein conjugated *E. coli* or fluorescein conjugated *S. aureus* cells in PBS (*see* Subheading 2.1) in the tube containing the hemolymph sample diluted in PBS (*see* **Note 2**).
5. Gently mix the suspension and incubate at room temperature, in the dark, 10 min for phagocytosis assay with *E. coli*, and 30 min for phagocytosis assay with *S. aureus*. During incubation, gently mix the tube every 5 min.

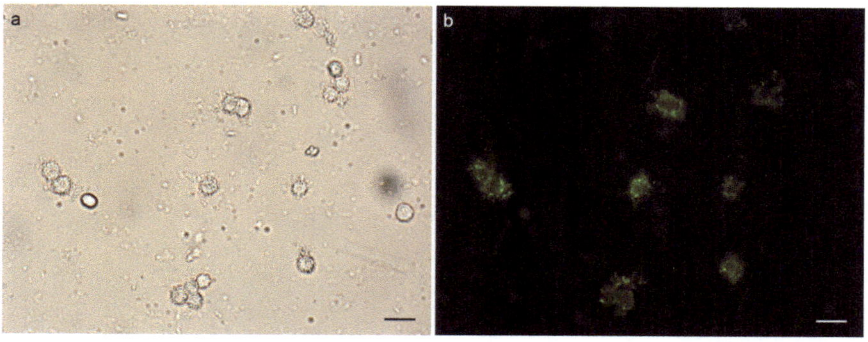

Fig. 1 Brightfield (**a**) and fluorescence (**b**) images of hemocytes extracted from *Spodoptera littoralis* larvae and incubated with fluorescently labeled *Escherichia coli*. The phagocytic hemocytes show a clear fluorescent signal (**b**). Bars, 15 μm. Source: Becchimanzi et al.

6. At the end of the incubation, gently mix the tube and transfer 10 μL of the cell suspension into a hemocytometer chamber.
7. Count the total number of hemocytes and the number of hemocytes showing internalized fluorescent bacteria (Fig. 1), using the fluorescence microscope at 400× magnification, and calculate the percentage of phagocytic hemocytes with the formula: phagocytic hemocytes (%) = (fluorescent hemocytes/ total hemocytes) × 100.

3.2 In Vivo Nodulation of Bacteria

1. Anaesthetize *S. littoralis* larvae using CO_2 for 2–3 min or by chilling on ice for 30 min.
2. Surface sterilize the experimental larva by immersion in 70% ethanol (5 s), which is then rinsed in sterile distilled water, and let it dry on a piece of autoclaved filter paper.
3. Before each injection, wash the Hamilton syringe with sequential rinses of sterile distilled water, 70% ethanol and sterile PBS.
4. Take 10 μL of 2×10^6/mL suspension of *E. coli* cells with the Hamilton syringe (*see* **Note 3**).
5. Place the larva on a parafilm piece, with the back facing upward.
6. Inject 10 μL of *E. coli* cells suspension through the neck membrane (*see* **Note 4**).
7. Wait a few seconds and, using forceps, gently transfer the larva into a vial containing a piece of artificial diet. Close the vial with cotton wool and place it in a climatic chamber at optimal rearing conditions.
8. 18 h after bacteria injection, anaesthetize the experimental larvae as described in **step 1**.
9. Place the larva on parafilm with the legs facing upward.

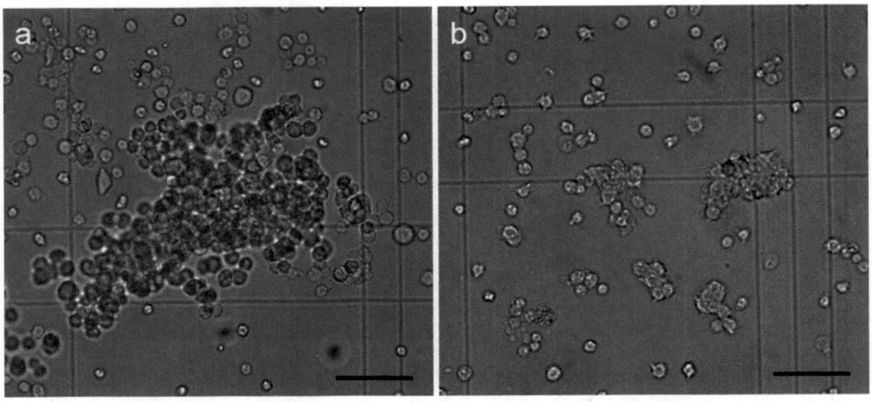

Fig. 2 Brightfield images of nodules isolated from *Spodoptera littoralis* larvae injected with *Escherichia coli*. In (**a**) and (**b**) nodules of different size and melanization degree are present. Bars, 50 μm. Source: Becchimanzi et al.

10. Cut a larval thoracic leg, and gently squeeze the larval body to collect the exuding hemolymph with a micropipette (range volume to 50 μL).
11. Transfer the collected hemolymph into a microcentrifuge tube (1.5 mL), containing ice-cold MEAD anticoagulant buffer (1:1 v/v).
12. Gently stir the tube and transfer 10 μL of the cell suspension into a hemocytometer chamber using a micropipette.
13. Count the nodules (Fig. 2) using a cell counter and calculate the number of nodules for μL of hemolymph with the appropriate formula for the hemocytometer chamber used (*see* **Note 5**).

3.3 In Vivo Encapsulation of Chromatography Beads

1. Anaesthetize *S. littoralis* larvae using CO_2 for 2–3 min or by chilling on ice for 30 min.
2. Surface sterilize the experimental larva by immersion in 70% ethanol (5 s), which is then rinsed in sterile distilled water, and let it dry on a piece of autoclaved filter paper.
3. Before each injection, wash the Hamilton syringe with sequential rinses of sterile distilled water, 70% ethanol and sterile PBS.
4. Mix gently the tube containing a suspension in PBS of beads, selected for adequate size using a cell strainer filter (*see* **Note 6**), and take 10 μL with the Hamilton syringe, which contain 20–25 beads (*see* **Note 7**).
5. Place the larvae on a parafilm piece, with the back facing upward.
6. Intrahemocelic injection of 10 μL of PBS, containing chromatography beads suspension, is performed through the neck

membrane (*see* **Note 4**). Gently remove the syringe needle (*see* **Note 8**).

7. Wait a few seconds and, using forceps, gently transfer the larva into a vial containing a piece of artificial diet. Close the vial with cotton wool, and place it in a climatic chamber at optimal rearing conditions.
8. After 24 h, anaesthetize the experimental larvae as described above, place them in a drop (2 mL) of MEAD solution in a Petri dish, and dissect the larval body under the stereomicroscope, using microscissors (*see* **Note 9**).
9. Gently remove the larval gut using a forceps and wash it in a fresh drop of MEAD solution; using forceps scrub the carcass of the larva into the MEAD drop, remove and put it in a fresh drop of MEAD solution. This step should be repeated, to recover about 80% of injected beads.
10. Using a stereomicroscope, recover the chromatography beads present in every drop (Fig. 3a), with a micropipette (range volume to 50 μL), and transfer the collected beads into a fresh drop of MEAD.
11. Transfer the beads in a well of a 96-well cell culture plate containing 200 μL of PBS.
12. Observe the beads using an inverted microscope to evaluate their encapsulation rate. The beads encapsulation index is calculated with the formula: [Σ (encapsulation degree × total beads of this degree)/ total beads × 4] × 100) that takes into account both the encapsulation degree of each recovered bead (0—no cells adherent to the beads, 1—up to 10 adherent cells, 2—more than 10 adherent cells but no complete layer

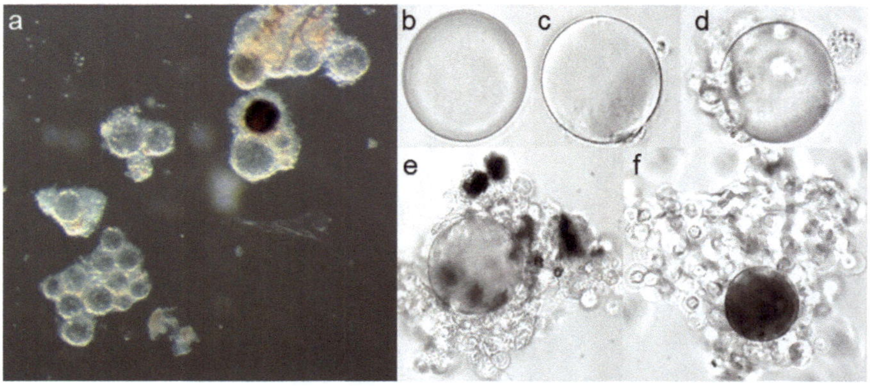

Fig. 3 In (**a**) beads isolated form *Spodoptera littoralis* larvae as appear under stereomicroscope observation. On the right panel, pictures of beads, observed under inverted microscope, showing different encapsulation levels: (**b**) 0 (no cells adherent to the beads); (**c**) 1 (up to 10 adherent cells); (**d**) 2 (more than 10 adherent cells but with less than a complete layer); (**e**) 3 (one or more complete layers without melanization); (**f**) 4 (one or more complete layers with melanization). Source: Becchimanzi et al.

around the bead, 3—one or more complete layers without melanization, 4—one or more complete layers with melanization) and the relative abundance of beads with a given encapsulation degree (Fig. 3b–f) [57].

3.4 In Vivo Encapsulation of Nylon Thread

1. Perform surface sterilization of forceps, scalpel, and needles in 70% ethanol.
2. Rub the nylon fishing thread (*see* **Note 10**) with fine sand paper in order to generate a rough surface, which facilitates hemocyte adhesion (*see* **Note 11**).
3. Use the scalpel to cut nylon thread in 2 mm pieces, with a sharp tip for an easier penetration in the larval cuticle.
4. Expose the nylon threads to a UV light source, under a horizontal laminar flow hood for 30 min.
5. Anaesthetize *S. littoralis* larvae using CO_2 for 2–3 min or by chilling on ice for 30 min.
6. Use a needle to make a hole through the cuticle at the base of the last pair of abdominal legs and insert the nylon thread in a way that does not impair the movements of the larva and prevents its shedding (Fig. 4a).
7. Keep each larva individually in cylindrical plastic vials capped with cotton wool with a piece of artificial diet, and maintain under optimal rearing conditions.
8. After 24 h, anesthetize the larvae as described at **step 5**.
9. Gently remove the implant with the forceps and place it on a microscope glass slide.

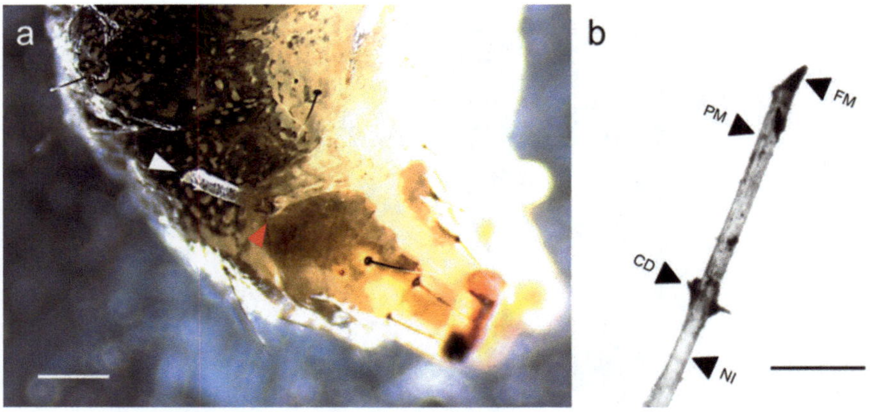

Fig. 4 Nylon thread is inserted in one of the last abdominal prolegs of *Spodoptera littoralis* larva (**a**). The white arrow indicates the external part and the orientation of the inserted nylon thread. Red arrow indicates the site of insertion through the cuticle at the base of the prolegs. Bar, 1 mm. Nylon thread is then removed from the larva after 24 h (**b**). *NI* non-inserted part, *CD* clotting and cuticle debris excluded from the subsequent image analysis, *PM* partially melanized portion, *FM* fully melanized portion. Bar, 500 μm. Source: Becchimanzi et al.

10. Use the digital camera mounted on the stereomicroscope to take at least two photographs of each implant, from different sides, at the same conditions of magnification, focus, and brightness.
11. Open the files with an image analysis software (e.g., GIMP, ImageJ, Photoshop, etc.).
12. Convert the image in black and white (Fig. 4b).
13. Cut the area of interest selecting only the implant part inserted into the insect body (*see* **Note 12**).
14. Use the Magic Wand tool to select the implant.
15. Open the Histogram and take note of the implant pixel number (IPN).
16. Select by color range only hemocyte-free parts on the implant (usually the brighter ones). Take note of the hemocyte-free pixel number (HFPN) and, then, convert the selection into white (*see* **Note 13**).
17. Reselect the implant as in **step 13** and take note of the average gray value (AGV), which represents the mean degree of gray intensity of the pixels in the implant area and consists in a numerical reading ranging from 0 for black to 255 for white (*see* **Note 14**).
18. Calculate the average of the values from the pictures of the same implant and insert them in the following formulae to obtain encapsulation and melanization indexes: nylon encapsulation index (%) = [(IPN − HFPN)/IPN] × 100; melanization index (%) = [1 − AGV/255] × 100.

4 Notes

1. The encapsulation efficiency of hemocytes from diverse lepidopteran species differs depending on beads material. Indeed, the capsule formation around foreign objects, such as chromatography beads, is significantly affected by surface features like charge, hydrophobicity, matrix composition, and the presence of functional groups [52, 58, 59]. Then, it is highly desirable to start with a preliminary encapsulation assay. For *S. littoralis* larvae, the selection of CM Sepharose® Fast Flow chromatography beads was done by screening for encapsulation (i) a number of beads made by different materials (Sephadex G 50 Fine, DEAE Sephadex A-25; CM-Sephadex, DEAE Sepharose, CM Sepharose, Dowex 50W) and (ii) for different time periods (12, 24, 48, 72 h).
2. The final concentration of bacterial cells in the hemolymph sample and the duration of incubation could strongly differ,

depending on the type of bacteria [29–33]. When switching to a different lepidopteran/bacterial species, these two key parameters must be preliminarily optimized. In order to assess hemocytes viability prior to proceed with the phagocytosis measurement, a viability test is performed using trypan blue staining. To do this, mix a hemolymph aliquot with 0.4% (w/v) trypan blue (2:1 v/v), and incubate for 5 min. Transfer 10 μL of sample into an hemocytometer chamber and count viable (clear cytoplasm) and dead cells (blue cytoplasm), under a light transmitted microscope. Calculate the percentage of viable cells with the formula: [1 – (number of blue cells/total cell number)] × 100. For optimal phagocytosis measurement, the percentage of viable cells should be higher than 95%.

3. To assess the nodulation response in vivo in a different lepidopteran species, it is essential to optimize the concentration of bacterial cells to be injected and the time interval between injection and dissection for each type of bacteria (*see* Subheading 1).
4. During the injection, the syringe should be held in a position parallel to the larval body, in order to prevent gut damage. Gently remove the syringe needle without applying any pressure on larval body, to avoid the leakage of hemolymph.
5. To avoid underestimation of the nodulation response, when an intense immune reaction gives rise to large aggregates of merging nodules (Fig. 2a) difficult to count separately, the number of original discrete nodules is obtained by doubling the number of nodule aggregates recorded, because the percentage of non-white pixels measured on the large aggregates is on average twice that measured on a bright-microscopy field containing discrete nodules and free hemocytes (Fig. 2b).
6. The selection of bead size range (in our case 45–70 μm) is performed by filtration through a cell strainer fitting to 50 mL Falcon tubes (70 μm). To do this, pipet 2 mL of the chromatography beads suspension into the strainer and collect the beads filtrating through the membrane in a 50 mL Falcon tube; wash three times the filter with 1 mL PBS. Repeat the filtration procedure using a new strainer until the filtrate contains beads of the size needed. Collect 1 mL of filtrate, and after centrifugation for 5 min at 14,000 × *g*, discard the supernatant and resuspend the pelleted beads in 1 mL PBS.
7. In our encapsulation assays, we inject 10 μL of PBS solution containing 20–25 chromatography beads. To obtain this bead concentration, 10 μL of the original filtrate are collected with the same syringe used for the intrahemocoelic injection, serially diluted as needed and the beads counted under a stereomicroscope. Once obtained the desired concentration, the

bead solution is aliquoted (100 μL in sterile microcentrifuge tubes) and autoclaved.

8. Gently remove the syringe needle from the larva. Holding the needle inside the larval body 5–10 s after injection limits the leakage of hemolymph.
9. Larva dissection is usually performed by cutting the head and then by cutting ventrally the neck membrane (as close as possible to the cephalic capsule), avoiding gut damage and contamination with gut contents of the hemolymph exuding into the MEAD drop. Then, the rest of the integument is cut lengthwise carefully, preventing gut damage.
10. The implant of nylon threads has been widely used to measure encapsulation and melanization response in insects [60–65] and appears to be a reliable method to assess the ability to encapsulate a real pathogen/parasite [66].
11. The rough surface of the implant is of pivotal importance to allow optimal hemocyte adhesion and stability of the capsule, preventing its shedding when the implant is removed from the experimental larva.
12. The site of insertion into the tegument of the nylon thread has to be excluded from the image part to be considered, as it can be irregularly covered by clotting clumps and cuticle pieces, which can negatively influence the image analysis.
13. Converting hemocyte-free pixels into white is essential for calculating melanization as a percentage on the whole range of gray scale values.
14. The melanization index (a percentage) is calculated as $(1 - AGV/255) \times 100$, where AGV represents the mean degree of gray intensity (a numerical reading ranging from 0 for black to 255 for white) of the pixels in the implant area. However, portions of the implant, which are not inserted in the insect, hemocyte-free and/or non-melanized, display a gray value different from 255 (i.e., total white), usually around 180.

References

1. Milutinović B, Peuß R, Ferro K et al (2016) Immune priming in arthropods: an update focusing on the red flour beetle. Zoology 119:254–261
2. Cooper D, Eleftherianos I (2017) Memory and specificity in the insect immune system: current perspectives and future challenges. Front Immunol 8:539
3. Lavine MD, Strand MR (2002) Insect hemocytes and their role in immunity. Insect Biochem Mol Biol 32:1295–1309
4. Tsakas S, Marmaras VJ (2010) Insect immunity and its signalling: an overview. Invertebr Surviv J 7:228–238
5. Krautz R, Arefin B, Theopold U (2014) Damage signals in the insect immune response. Front Plant Sci 5:342
6. Rosales C (2017) Cellular and molecular mechanisms of insect immunity. In: Insect physiology and ecology. InTechOpen, Croatia
7. Mylonakis E, Podsiadlowski L, Muhammed M et al (2016) Diversity, evolution and

medical applications of insect antimicrobial peptides. Philos Trans R Soc Lond B Biol Sci 371:20150290

8. Wu Q, Patočka J, Kuča K (2018) Insect antimicrobial peptides, a mini review. Toxins 10:461
9. Bogdan C, Röllinghoff M, Diefenbach A (2000) The role of nitric oxide in innate immunity. Immunol Rev 173:17–26
10. Vass E, Nappi AJ (2001) Fruit fly immunity. BioScience 51:529–535
11. González-Santoyo I, Córdoba-Aguilar A (2012) Phenoloxidase: a key component of the insect immune system. Entomol Exp Appl 142:1–16
12. Lu A, Zhang Q, Zhang J et al (2014) Insect prophenoloxidase: the view beyond immunity. Front Physiol 5:252
13. Sheehan G, Garvey A, Croke M et al (2018) Innate humoral immune defences in mammals and insects: the same, with differences? Virulence 9:1625–1639
14. Strand MR, Pech LL (1995) Immunological basis for compatibility in parasitoid-host relationships. Annu Rev Entomol 40:31–56
15. Schmidt O, Theopold U, Strand M (2001) Innate immunity and its evasion and suppression by hymenopteran endoparasitoids. BioEssays 23:344–351
16. Nappi AJ, Kohler L, Mastore M (2004) Signaling pathways implicated in the cellular innate immune responses of *Drosophila*. Invertebr Surviv J 1:5–33
17. Mavrouli MD, Tsakas S, Theodorou GL et al (2005) MAP kinases mediate phagocytosis and melanization via prophenoloxidase activation in medfly hemocytes. Biochim Biophys Acta BBA-Mol Cell Res 1744:145–156
18. Stanley DW, Miller JS (2006) Eicosanoid actions in insect cellular immune functions. Entomol Exp Appl 119:1–13
19. Sideri M, Tsakas S, Markoutsa E et al (2008) Innate immunity in insects: surface-associated dopa decarboxylase-dependent pathways regulate phagocytosis, nodulation and melanization in medfly haemocytes. Immunology 123:528–537
20. Marmaras VJ, Lampropoulou M (2009) Regulators and signalling in insect haemocyte immunity. Cell Signal 21:186–195
21. Lamprou I, Tsakas S, Theodorou GL et al (2005) Uptake of LPS/E. coli/latex beads via distinct signalling pathways in medfly hemocytes: the role of MAP kinases activation and protein secretion. Biochim Biophys Acta BBA-Mol Cell Res 1744:1–10
22. Jiang H, Vilcinskas A, Kanost MR (2010) Immunity in lepidopteran insects. In: Invertebrate immunity. Springer, New York, pp 181–204
23. da Silva CC, Dunphy GB, Rau ME (2000) Interaction of *Xenorhabdus nematophilus* (Enterobacteriaceae) with the antimicrobial defenses of the house cricket, *Acheta domesticus*. J Invertebr Pathol 76:285–292
24. Hernández S, Lanz H, Rodríguez MH et al (1999) Morphological and cytochemical characterization of female *Anopheles albimanus* (Diptera: Culicidae) hemocytes. J Med Entomol 36:426–434
25. Nazario-Toole AE, Wu LP (2017) Phagocytosis in insect immunity. In: Advances in insect physiology. Elsevier, London, pp 35–82
26. Rosales C (2011) Phagocytosis, a cellular immune response in insects. Invertebr Surviv J 8:109–131
27. Yokoo S, Gotz P, Tojo S (1995) Phagocytic activities of haemocytes separated by two simple methods from larvae of two lepidopteran species, *Agrotis segetum* and *Galleria mellonella*. Appl Entomol Zool 30:343–350
28. Oliver JD, Loy JD, Parikh G et al (2011) Comparative analysis of hemocyte phagocytosis between six species of arthropods as measured by flow cytometry. J Invertebr Pathol 108:126–130
29. Lamprou I, Mamali I, Dallas K et al (2007) Distinct signalling pathways promote phagocytosis of bacteria, latex beads and lipopolysaccharide in medfly haemocytes. Immunology 121:314–327
30. Levashina EA, Moita LF, Blandin S et al (2001) Conserved role of a complement-like protein in phagocytosis revealed by dsRNA knockout in cultured cells of the mosquito, *Anopheles gambiae*. Cell 104:709–718
31. Rämet M, Manfruelli P, Pearson A et al (2002) Functional genomic analysis of phagocytosis and identification of a *Drosophila* receptor for E. coli. Nature 416:644
32. García-García E, García-García PL, Rosales C (2009) An fMLP receptor is involved in activation of phagocytosis by hemocytes from specific insect species. Dev Comp Immunol 33:728–739
33. Hillyer JF, Schmidt SL, Christensen BM (2003) Rapid phagocytosis and melanization of bacteria and *Plasmodium sporozoites* by hemocytes of the mosquito *Aedes aegypti*. J Parasitol 89:62–69
34. Elrod-Erickson M, Mishra S, Schneider D (2000) Interactions between the cellular and humoral immune responses in *Drosophila*. Curr Biol 10:781–784
35. Meister M (2004) Blood cells of Drosophila: cell lineages and role in host defence. Curr Opin Immunol 16:10–15
36. Castillo JC, Robertson AE, Strand MR (2006) Characterization of hemocytes from the mos-

quitoes *Anopheles gambiae* and *Aedes aegypti*. Insect Biochem Mol Biol 36:891–903
37. Lemaitre B, Hoffmann J (2007) The host defense of *Drosophila melanogaster*. Annu Rev Immunol 25:697–743
38. Vogelweith F, Moret Y, Monceau K et al (2016) The relative abundance of hemocyte types in a polyphagous moth larva depends on diet. J Insect Physiol 88:33–39
39. Ratcliffe NA, Walters JB (1983) Studies on the in vivo cellular reactions of insects: clearance of pathogenic and non-pathogenic bacteria in *Galleria mellonella* larvae. J Insect Physiol 29:407–415
40. Satyavathi VV, Minz A, Nagaraju J (2014) Nodulation: an unexplored cellular defense mechanism in insects. Cell Signal 26:1753–1763
41. Dubovskiy IM, Kryukova NA, Glupov VV et al (2016) Encapsulation and nodulation in insects. Invertebr Surviv J 13:229–246
42. Ratcliffe NA, Gagen SJ (1977) Studies on the in vivo cellular reactions of insects: an ultrastructural analysis of nodule formation in *Galleria mellonella*. Tissue Cell 9:73–85
43. Ling E, Yu X-Q (2006) Cellular encapsulation and melanization are enhanced by immulectins, pattern recognition receptors from the tobacco hornworm *Manduca sexta*. Dev Comp Immunol 30:289–299
44. Nappi A, Poirié M, Carton Y (2009) The role of melanization and cytotoxic by-products in the cellular immune responses of *Drosophila* against parasitic wasps. Adv Parasitol 70:99–121
45. Nappi AJ, Vass E, Frey F et al (1995) Superoxide anion generation in *Drosophila* during melanotic encapsulation of parasites. Eur J Cell Biol 68:450–456
46. Nappi AJ, Ottaviani E (2000) Cytotoxicity and cytotoxic molecules in invertebrates. BioEssays 22:469–480
47. Carton Y, Nappi AJ, Poirie M (2005) Genetics of anti-parasite resistance in invertebrates. Dev Comp Immunol 29:9–32
48. Dubovskiy IM, Grizanova EV, Chertkova EA et al (2010) Generation of reactive oxygen species and activity of antioxidants in hemolymph of the moth larvae *Galleria mellonella* (L.) (Lepidoptera: Pyralidae) at development of the process of encapsulation. J Evol Biochem Physiol 46:35–43
49. Pech LL, Strand MR (1996) Granular cells are required for encapsulation of foreign targets by insect haemocytes. J Cell Sci 109:2053–2060
50. Hillyer JF (2016) Insect immunology and hematopoiesis. Dev Comp Immunol 58:102–118
51. Kanost MR, Jiang H, Yu X-Q (2004) Innate immune responses of a lepidopteran insect, *Manduca sexta*. Immunol Rev 198:97–105
52. Lackie AM (1988) Haemocyte behaviour. In: Advances in insect physiology. Elsevier, London, pp 85–178
53. Meister M, Lagueux M (2003) *Drosophila* blood cells. Cell Microbiol 5:573–580
54. Ribeiro C, Brehélin M (2006) Insect haemocytes: what type of cell is that? J Insect Physiol 52:417–429
55. Strand MR (2008) The insect cellular immune response. Insect Sci 15:1–14
56. Di Lelio I, Varricchio P, Di Prisco G et al (2014) Functional analysis of an immune gene of *Spodoptera littoralis* by RNAi. J Insect Physiol 64:90–97
57. Di Lelio I, Illiano A, Astarita F et al (2019) Evolution of an insect immune barrier through horizontal gene transfer mediated by a parasitic wasp. Plos Genet. https://doi.org/10.1371/journal.pgen.1007998
58. Lavine MD, Strand MR (2001) Surface characteristics of foreign targets that elicit an encapsulation response by the moth *Pseudoplusia includens*. J Insect Physiol 47:965–974
59. Gorman MJ, Schwartz AM, Paskewitz SM (1998) The role of surface characteristics in eliciting humoral encapsulation of foreign bodies in *Plasmodium*-refractory and -susceptible strains of *Anopheles gambiae*. J Insect Physiol 44:947–954
60. Kivleniece I, Krams I, Daukšte J et al (2010) Sexual attractiveness of immune-challenged male mealworm beetles suggests terminal investment in reproduction. Anim Behav 80:1015–1021
61. Koskimäki J, Rantala MJ, Taskinen J et al (2004) Immunocompetence and resource holding potential in the damselfly, *Calopteryx virgo* L. Behav Ecol 15:169–173
62. Krams I, Daukste J, Kivleniece I et al (2013) Previous encapsulation response enhances within individual protection against fungal parasite in the mealworm beetle *Tenebrio molitor*. Insect Sci 20:771–777
63. Rantala MJ, Jokinen I, Kortet R et al (2002) Do pheromones reveal male immunocompetence? Proc Biol Sci 269:1681–1685
64. Rantala MJ, Kortet R, Kotiaho JS et al (2003) Condition dependence of pheromones and immune function in the grain beetle *Tenebrio molitor*. Funct Ecol 17:534–540
65. Vainio L, Hakkarainen H, Rantala MJ et al (2004) Individual variation in immune function in the ant *Formica exsecta*; effects of the nest, body size and sex. Evol Ecol 18:75–84
66. Rantala MJ, Roff DA (2007) Inbreeding and extreme outbreeding cause sex differences in immune defence and life history traits in *Epirrita autumnata*. Heredity 98:329–336

Part III

Humoral Response: Melanization, Lysozyme, and Antimicrobial Peptides

Chapter 7

Detection of Enzyme Distribution, Expression, Activation, and Activity of Insect Prophenoloxidase

Kai Wu, Bing Yang, Jing Wang, Yufa Luo, Yuyang Ni, and Wuren Huang

Abstract

Melanization is an important defense mechanism of innate immunity in insects. Prophenoloxidase (PPO) plays a key role in the process of melanization and is expressed in the form of an inactive zymogen. Upon activation, phenoloxidase (PO) oxidizes substrates leading to the formation of melanin through a series of reactions. To study the melanization reaction in insects, it is necessary to identify which tissues and cells express PPO and to understand the mechanism of action of PPO and how it interacts with other proteins in and outside the cell. PO activity assays and activation methods are necessary tools to probe the role and function of PPO in insect melanization. Here we describe methods of easy experimental setup to detect PPO expression in insect tissues and cells, prokaryotic expression of PPO, PPO activation, and enzyme activity. The use of these methods may be helpful to scientists who are engaged in insect PPO and melanization research.

Key words Activity, Distribution, Prophenoloxidase

1 Introduction

Prophenoloxidase (PPO) is an important immunoprotein in insects and a key melanization-inducing enzyme. Melanization plays an important role in blood coagulation, wound healing, and killing pathogens [1–4]. As it is linked to the humoral and cellular immunity of insects, melanization is also associated with the Toll pathway and various immune factors [5, 6]. PPO and melanization have been an important topic in insect immunity.

Insect PPO has been studied for over 100 years [7]. As a type 3 copper-containing protein, the PPO family exists in almost all living organisms and comprises other enzymes such as polyphenol oxidase, tyrosinase, and hemocyanin [8, 9]. Insect PPO is expressed in hemocytes but contains no signal peptide, and thus its mechanism of secretion remains unclear. Some researchers believe that PPO secretion occurs during cell rupture; transgenic *Drosophila* expressing PPO-GFP fusion protein may help to prove subcellular

Federica Sandrelli and Gianluca Tettamanti (eds.), *Immunity in Insects*, Springer Protocols Handbooks,
https://doi.org/10.1007/978-1-0716-0259-1_7,

localization and transport of PPO in the cell [10]. Many kinds of PPOs in different species have been identified and their structures resolved, such as *Manduca sexta* and *Marsupenaeus japonicus* [8, 11]. In addition to hemocytes, PPO also exists in the foregut, hindgut, and molting fluid of insects [12–14]. PPO in the foregut can detoxify phenolic compounds in food forming low-toxicity intermediates, which can be transported to the hindgut through hemolymph. In the hindgut, PPO can further oxidize the intermediates to form melanin, which helps kill fecal bacteria and maintain a clean environment [12, 14]. In molting fluid, PPO acts as a barrier against fungal infection [13, 15].

PPO has various activation and inhibition modes. However, it is generally believed that PPO exists in the form of an inactive zymogen, which can be cleaved and activated by serine proteases or inhibited by the serine protease inhibitor serpin [6, 7, 16, 17]. At present, the mechanism of PPO activation and regulation in insect digestive tracts is unknown, whereas the PPO pathway in hemolymph is well understood. However, close attention should be paid to phenoloxidase (PO) activity as it can be activated in a variety of ways and the same substrate can be oxidized to form distinct intermediates by PO [18]. In addition to the various PPO activation pathways within individual insects, the number of PPOs in different insect species also varies. To date, the melanization reaction pathway in hemolymph is not well understood. Therefore, before studying PPO and melanization in insects, it is necessary to know which type of tissues or cells express PPO. To study the kinetic characteristics of PO in vivo or in vitro, the expression and purification of PPO protein are required. Monitoring of melanization reaction and detection of melanin are also important for studying the function of PPO. Previous studies have shown that, by suppressing immune response or reducing PPO expression via RNAi, insects are more likely to be killed by pathogens [19]. Moreover, in *PPO1* and *PPO2* deletion mutants, *Drosophila* larvae and adults are more susceptible to the toxicity of plant metabolites, such as phenolics, added to the food [12]. The impairment of insect immune response, by inhibiting PPO activation or inducing abnormal melanization in the insect, could represent an interesting tool for pest control. Consequently, it is mandatory to establish appropriate methods to evaluate PO activity and to study melanization in insects.

In this chapter, we review detection methods of PPO in insect tissues and cells, in vitro expression and purification of PPO, activation, detection methods for PPO activity, and methods for detecting intermediates in the melanization process.

2 Materials

2.1 Insect Dissection

1. Living insects.
2. 0.85% $NaCl_{(aq)}$ (w/v%), phosphate-buffered saline (PBS; 8 g NaCl, 0.2 g KCl, 1.42 g Na_2HPO_4, and 0.27 g KH_2PO_4 dissolved in 800 mL ddH_2O using NaOH to adjust to pH 7.4, and made up to 1 L), 30% $ethanol_{(aq)}$ (v/v%) containing 2 mM L-3,4-dihydroxyphenylalanine (L-DOPA) (30 mL ethanol, 70 mL ddH_2O, and 39.4 mg L-DOPA [Abcam, ab120573]), 30% $ethanol_{(aq)}$ (v/v%) containing 10 mM dopamine (30 mL ethanol, 70 mL ddH_2O, and 189.64 mg dopamine [Sigma-Aldrich, H8502]).
3. Basic anatomical instruments such as Petri dishes, tweezers, dissecting scissors, and dissecting discs.
4. Stereoscope (Olympus, SZ51) and camera (Nikon, D7200) for snap picture. Centrifuge (Eppendorf, 5424R), microscope (Olympus, BX53), and inverted microscope (Olympus, CKX41) for cell collection and observation.

2.2 rPPO1 Purification and Color Reaction

1. *Escherichia coli* expressing strain: *PPO1* recombinant *E. coli* (BL21) (transformed with pET28a-PPO1 plasmid). *PPO1*, *Drosophila melanogaster PPO1* gene (CG5779).
2. Fresh insect (*Bombyx mori* larvae) hemolymph.
3. Culture medium: Lysogeny Broth (LB) medium.
4. Isopropyl β-D-1-thiogalactopyranoside (IPTG): 1 M.
5. Binding buffer: 10 mM Tris–HCl, pH 7.0.
6. Wash buffers: 10, 20, and 50 mM imidazole in 10 mM Tris–HCl, pH 7.4, containing 100 mM NaCl.
7. Elution buffer: 100 mM imidazole in 10 mM Tris–HCl, pH 7.4, containing 100 mM NaCl.
8. Dye reagent concentrate (Bio-Rad, 500-0006), Ni-NTA agarose (Qiagen, 30210).
9. Reagents (12% acrylamide gel; 1× running buffer: 3.03 g Tris base, 14.4 g glycine, and 1 g sodium dodecyl sulfate [SDS], in 1 L ddH_2O) are needed to perform SDS-polyacrylamide gel electrophoresis (PAGE).
10. For color reaction: 30% $ethanol_{(aq)}$ (v/v%), 10 mM dopamine, 5 mM $CuSO_4$ or $CuCl_2$, PBS.
11. Ultrasonic cell disruption device (SCIENTZ, 950E) is used to disrupt *E. coli* cells and a centrifuge (Eppendorf, 5810R; Eppendorf, 5424R) to collect the bacteria. Electrophoresis apparatus (BIO-RAD, PowerPac™ HC, 164-5052).

2.3 PO Activity Assay

To activate PPO, use 30% ethanol$_{(aq)}$ (v/v%). Serine proteases (AMM1) or α-chymotrypsin can also cleave PPO into PO [17, 20]. To detect PO activity, use *N*-phenylthiourea (PTU; Sigma P7629) to inhibit PO activity as a control, 10 mM dopamine as the substrate. 96-well microtiter plates and a microplate spectrophotometer (Varioskan™ Flash Multimode Reader) are used to detect PO activity. A unit of enzyme activity is defined as $\Delta A\lambda$/min = 0.001 (λ, 490 nm).

2.4 Native Gel

1. 8% Native gel: 1.75 mL ddH_2O, 1.3 mL 30% acrylamide$_{(aq)}$, 1.9 mL 1.5 M Tris–HCl (pH 8.8), 50 μL 10% ammonium persulfate$_{(aq)}$, and 3 μL tetramethylethylenediamine (TEMED) are used to prepare 8% resolving gels; 1.4 mL ddH_2O, 0.32 mL 30% acrylamide$_{(aq)}$, 0.25 mL 1 M Tris–HCl (pH 6.8), 20 μL 10% ammonium persulfate$_{(aq)}$, and 2 μL TEMED are used to prepare 5% stacking gels (*see* **Note 1**).
2. Electrophoresis buffer: 3.03 g Tris base and 14.4 g glycine; make up to 1 L with ddH_2O.
3. 5× native loading buffer: 1.25 mL 1 M Tris–HCl (pH 6.8), 25 mg bromophenol blue, and 2.5 mL glycerol; make up to 5 mL with ddH_2O.
4. Staining buffer: 10 mM dopamine$_{(aq)}$ (for PO) or 30% ethanol$_{(aq)}$ (v/v%) containing 10 mM dopamine (for PPO); 30% ethanol$_{(aq)}$ (v/v%) containing 1 mM gallic acid, 0.5 mM tannic acid, 0.36 mM gossypol, 0.2 mM quercetin, and 0.2 mM chlorogenic acid separately.

2.5 High-Performance Liquid Chromatography-Mass Spectrometry (HPLC-MS)

1. 0.22 μm filter membrane (Millex® GP Filter Unit 0.22 μm) is used to remove particles.
2. An accurate-mass LC-1200/QTOF6520A system (Agilent Technologies) is used for mass spectrometric analysis; reversed-phase column: 2.1-mm diameter, 30-mm length, 3.5-μm particle size (Agilent Zorbax SB-C18).
3. Mobile phase: A, 0.1% formic acid in water; B, 0.1% formic acid in methanol.
4. Mobile phase gradient of 5–15% B (0–15 min), 15% B (15–19 min), and 15–5% B (19–20 min) at a flow rate of 0.2 mL/min. Set the diode array detector at 254 and 280 nm to detect L-DOPA and intermediates. The mass spectra are collected using a mass range of 100–1700 Da with a nebulizer pressure of 40 psig and N_2 gas (350 °C) at a flow rate of 9 mL/min. Set voltages as follows: electrospray ionization Vcap, 3500 V; fragmentation voltage, 160 V; skimmer voltage, 65 V; and Oct RF Vpp, 750 V [18].
5. Software: MassHunter Qualitative Analysis software (version B.03; Agilent Scientific Co. Ltd.) and MassProfiler software (version B.02; Agilent Scientific Co. Ltd.).

3 Methods

3.1 Detection of PPO in Insect Tissues

1. Dissect fifth instar silkworm larvae to obtain the intestinal tract, including the foregut, midgut, and hindgut (*see* **Note 2**).
2. Submerge the whole gut in 0.85% $NaCl_{(aq)}$ (w/v%) solution and wash three times to remove the gut content and hemolymph (*see* **Note 3**).
3. Submerge the whole gut in 30% $ethanol_{(aq)}$ (v/v%) containing 10 mM dopamine at room temperature.
4. Stain the gut until the foregut and hindgut are darkened (several hours) (*see* **Note 4**), and then take pictures of the gut by camera or stereoscopy (Fig. 1a).

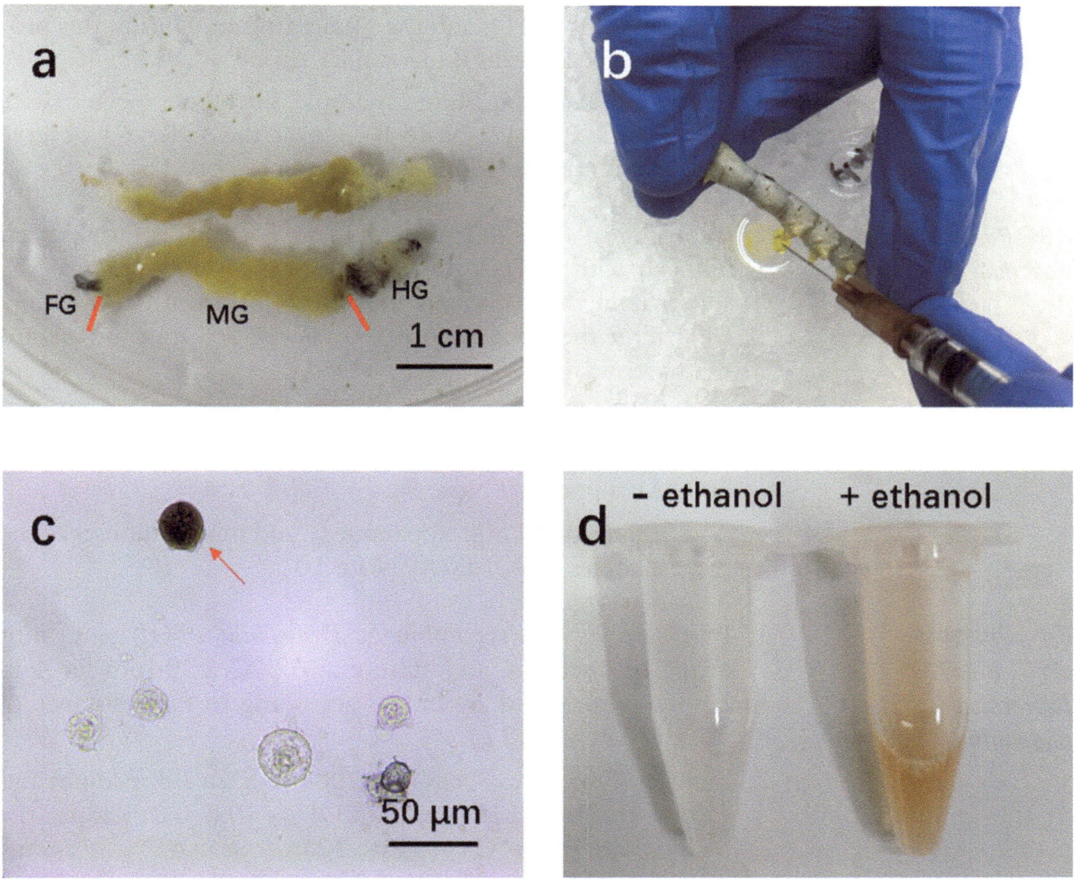

Fig. 1 (**a**) The foregut and hindgut contain prophenoloxidase (PPO). The gut stained with 30% $ethanol_{(aq)}$ contains 10 mM dopamine; prior to staining, the whole gut was not black (upper). The foregut and hindgut stain black when PPO is activated. *FG* foregut, *MG* midgut, *HG* hindgut. (**b**) Hemolymph collection method. (**c**) PPO-expressing hemocytes can be stained black. The arrow shows the melanized cell. (**d**) Color reaction. PPO was not activated in the left tube

3.2 Cell Detection of PPO Expression

1. Wash the surface of the insect with water and dry using tissue. Restrain the head and tail of silkworm larvae (e.g., Lepidoptera larvae) with fingers (Fig. 1b) (*see* **Note 5**).
2. Puncture (by using a needle) the first proleg of the 5 fifth instar silkworm larvae (V-3), and collect hemolymph with 1.5-mL centrifuge tubes. Place the centrifuge tubes on ice until 200 μL hemolymph centrifugation at 4 °C for 3 min at 2500 × *g*.
3. Remove the plasma with a pipette and suspend the hemocytes gently in 100 μL 0.85% $NaCl_{(aq)}$ solution or PBS buffer (*see* **Note 6**). Place a 50 μL suspension of hemocytes onto a slide and leave for 10 min to allow cells to adhere to the slide (*see* **Note 7**). Use inverted microscopy to verify whether the cells are present on the slide.
4. Remove the buffer gently from the slide. Fix the cells with 50 μL 35% $ethanol_{(aq)}$ (v/v%) containing 10 mM dopamine, which activates PPO, and then stain 40 min. During this process, observe the cells under a microscope.
5. Remove the staining solution and wash the hemocytes twice with 50 μL PBS (*see* **Note 8**). Cover the slide and observe which cells are stained black under microscope (Fig. 1c).

3.3 PO Color Reaction

1. Centrifuge 1 mL PPO1 recombinant *E. coli* cells (it had been induced with IPTG) at 10,000 × *g* for 1 min, discard the supernatant, and suspend the pellets in 1 mL PBS buffer. Centrifuge the suspension at 10,000 × *g* for 1 min and discard the supernatant.
2. Add 45 μL 30% $ethanol_{(aq)}$ (v/v%) solution to activate PPO.
3. Add 5 μL 5 mM $CuSO_4$ to obtain a final concentration of 500 μM copper (*see* **Note 9**).
4. Add 500 μL of 10 mM $dopamine_{(aq)}$ and mix, and observe the color changes (Fig. 1d) (*see* **Note 10**).

3.4 Drosophila rPPO1 (CG5779) Expression and Purification

1. Inoculate *E. coli* (recombinant *Drosophila PPO1* gene) into 4 mL LB medium, and culture at 37 °C, 220 rpm for 4 h. Before culture, add antibiotics according to the resistance of plasmid (*see* **Note 11**).
2. Transfer 4 mL *E. coli* cells into 200 mL fresh LB medium and culture at 37 °C, 220 rpm until OD_{600} ~0.6–0.8. Reduce the temperature to 16 °C (*see* **Note 12**).
3. After reaching 16 °C, add 20 μL of 1 M IPTG (final concentration 0.1 mM IPTG) and induce the cells for 12 h or overnight.
4. Detect rPPO1 expression by the method described in Subheading 3.3.

5. Pour 50 mL of the medium into a 50-mL centrifuge tube and centrifuge at 4 °C for 10 min at 6000 × *g* to collect all bacterial cells (four times), and lastly discard the supernatant and add 20 mL of pre-cooled binding buffer (10 mM Tris–HCl buffer, pH 7.0). Suspend and centrifuge twice more, discard the supernatant, and add fresh buffer each time. Discard the supernatant.
6. Add 20 mL of binding buffer containing 100 mM NaCl buffer to suspend the bacteria and sonicate the cells 30 min at 0 °C. Ultrasonic power: 200 W. Each sonication for 10 s with a 20 s break in between (*see* **Note 13**).
7. Centrifuge at 4 °C, 12,000 × *g* for 10 min. Collect the supernatant and avoid precipitation (*see* **Note 14**).
8. Add 1 mL Ni-NTA agarose to a column and balance with binding buffer (*see* **Note 15**). Incubate the supernatant with Ni-NTA agarose at 4 °C for 40 min. Transfer the mixture to the column and drip freely; beads can stay in the column. Wash with binding buffer one time; use 3 × 10 mL washing buffer (Tris–HCl containing 10 mM imidazole) to elute undesired proteins (*see* **Note 16**).
9. Wash the column with 2 mL 20 mM imidazole Tris–HCl, 1 mL 30 mM imidazole Tris–HCl, and 1 mL 50 mM imidazole Tris–HCl sequentially. Finally, elute the column with 4–5 mL elution buffer. Collect the eluted fractions in centrifuge tubes (*see* **Note 17**). Mix the collected protein solution with glycerol in equal parts and stored at −80 °C.
10. Use SDS-PAGE (180 V, 45 min) and Coomassie staining assays to assess protein purity and quantity.

3.5 PO Activity Assay

1. Activate rPPO1 by mixing equal parts of PPO with 60% $ethanol_{(aq)}$. Adding copper ions is also necessary (*see* **Note 18**).
2. Mix 1 μg rPO1 with 100 μL 10 mM $dopamine_{(aq)}$; perform the experiments in triplicate. Control experiments are performed with 10 μL saturated PTU solution to inhibit PO activity, while 10 μL ddH_2O is added to the experimental groups to make the reaction system consistent with the control groups. Read absorbance (490 nm) at 0, 5, 10, and 15 min by using the microplate spectrophotometer. A unit of enzyme activity is defined as $\Delta A\lambda/\text{min} = 0.001$ (λ, 490 nm).

3.6 Detection of PPO by Native Gel Electrophoresis

1. Sonicate the samples (tissues and cells) in 200 μL PBS buffer and centrifuge at 4 °C at 10,000 × *g*.
2. Mix 100 μL of the supernatant with 25 μL of 5× native-loading buffer.
3. Pipet samples (20 μL, ~2 μg PPO) onto the wells of native-gel and electrophoresis at a constant 300 V for 1 h at 4 °C (*see* **Note 19**).

Fig. 2 rPPO1 in native gel electrophoresis stained with different substrates. Dopamine (10 mM), gallic acid (1 mM), tannic acid (0.5 mM), gossypol (0.36 mM), quercetin (0.2 mM), and chlorogenic acid (0.2 mM) prepared in 30% ethanol$_{(aq)}$ (v/v%) were also used as substrates to detect rPPO1 (2 μg) using native gel electrophoresis as described. The arrow shows the position of rPPO1

4. Stain the gel with 10 mM dopamine$_{(aq)}$ until the PO band appeared (Fig. 2). If PPO in the sample is not activated, then 30% ethanol$_{(aq)}$ containing 10 mM dopamine is used to stain (*see* **Note 20**).

3.7 Identification of Intermediates in the rPO1 Catalytic Reaction System (L-DOPA as Substrate)

1. Add 20 μL saturated PTU to a 200 μL rPO1 reaction system to stop the process of blackening. Centrifuge the samples at 4 °C for 3 min, 12,000 × *g*, and remove impurities by 0.22 μm membrane filtration. The filtrate is analyzed by HPLC-MS (*see* **Note 21**).
2. Inject the sample (10 μL) into the mass spectrometer (*see* **Note 22**) and perform HPLC-MS in negative mode.
3. Data analysis is carried using MassHunter Qualitative Analysis and MassProfiler software. First, open the Qualitative Analysis software, and then open multiple raw data files that need to be analyzed and compared. Click the list mode. To compare relative quantities, calculate the area of the peak after extract MS spectrum. To find compounds, select "Find by Molecular Feature" mode, set the peaks with height more than 1200 counts, and then run the program to find compounds.
4. After the program complete, export the results as CEF and CSV files.
5. Open the MassProfiler software, click the File menu bar, create a new project, and add experiment samples CEF file and control samples CEF file. Plot style can set in different colors.

6. Properties such as retention time, mass spectra, topology dot distribution, and indicator/molecular ion peaks are used to identify the specific compounds. Comparison of the different molecules with standard identifiers from mass spectra databases, mainly use METLIN and MassBank.

4 Notes

1. The thickness of native gels is 0.75 mm. Roughly 3.8 mL of 8% resolving gel mixture is allowed to solidify and ~2 mL of 5% stacking gel mixture is added. Place a comb at the top until the gel has set.
2. Sampling methods of other tissues, such as fat body, malpighian tubules, and tracheae, are similar. The tissues can also be stained separately; e.g., wing discs of silkworm larvae can also be stained black. This was proved to be the case after PPO in wing discs was released by hemocytes from the hematopoietic organs [21].
3. The dissection needs to be carried out swiftly as intestinal tissue degrades easily with time. This is particularly important for the midgut of last instar Lepidoptera larvae. PPO may also degrade over time, resulting in low activity and a poor dyeing effect.
4. The foregut and hindgut of carnivorous insects cannot be stained black with staining buffer because there is no PPO expression.
5. If the insect is small, under the stereomicroscope, tear open the insect body using two forceps to release hemocytes in a drop of 0.85% $NaCl_{(aq)}$.
6. Without removing plasma, residual PPO from plasma can cause melanization.
7. This experiment is carried out in a wet box; otherwise, the cell suspension will dry easily.
8. The background is too dirty without washing, affecting the quality of the picture.
9. Without Cu^{2+}, the recombinant PPO1 has no activity even when incubated with 30% $ethanol_{(aq)}$. The rPO1 activity can be inhibited if the concentration of Cu^{2+} is too high. However, some PPOs can self-activate, negating the need for copper or ethanol [20, 22].
10. The color reaction experiment of PO in plasma is similar. 10 μL insect hemolymph must be collected and incubated with 100 μL 10 mM $dopamine_{(aq)}$ to observe whether there is color reaction. If so, PPO in insect hemolymph is activated.

11. Add 4 μL of 50 mg/mL kanamycin to 4 mL of LB medium for plasmid is pET28a-rPPO1.
12. Lower temperature contributes to soluble-rPPO1 expression.
13. Cool the tube to 0 °C during sonication because heat is generated. Avoid centrifuge tube dislodging and bubble formation in the bacterial liquid when ice melt. Sonication is carried out until the bacterial liquid becomes transparent.
14. There are many undesired proteins in the precipitate, which might affect the purity of rPPO1.
15. Ensure there are no other proteins present. If so, elute with 1–1.5 M imidazole$_{(aq)}$ and balance the column.
16. Protein can be detected with dye reagent concentrate. If protein exists, the buffer turns blue. Detection method: 1:10 volume ratio, 5 μL dye reagent concentrate + 50 μL water + 5 μL eluent.
17. Generally, protein degradation can be reduced by cooling to 0 °C. However, protein precipitation can occur at 0 °C if the concentration of eluted proteins is high, which reduces the yield of purified PPO.
18. Purified PPO needs to be activated. The most commonly used method of activation is mixed with organic solvents such as ethanol or cetylpyridinium chloride. Other activation methods include the use of α-chymotrypsin and AMM1. Prior to activation, imidazole must be removed as it can make the substrates of L-DOPA or dopamine produce a nonenzymatic color reaction; it also cannot be inhibited by PTU.
19. If the PPO quantity in the sample is too low, use 1.5 mm gels with 50 μL samples.
20. In general, insect tissue samples do not need additional Cu^{2+} during staining. Purified rPPO1 requires Cu^{2+} to be added to the staining solution. This method can also detect PPO activation by other proteinases in vitro. The entire gel will be stained black if the stain is left for too long.
21. After filtration, the filtrate was cooled to 0 °C; the intermediate products are unstable and oxidize easily; therefore it is necessary to perform HPLC-MS as soon as possible.
22. It is better to have standard chemical compounds, such as 5,6-dihydroxyindole (DHI). If there is a standard, it can be used with HPLC/QqQ-MS (triple quadrupole mass spectrometer). However, to identify nonstandard substances, the use of nuclear magnetic resonance or full scan mass spectrometry is necessary to collect data and allow comparisons to be made with public databases.

Acknowledgments

This work was supported by the Project of Science and Technology of Jiangxi (20192BAB214009, 20171BAB216038), Shanghai Natural Science Foundation (19ZR1466500), National Natural Science Foundation of China (31872294, 31672360 and 31660611), Science and Technology Foundation of Educational Commission of Jiangxi Province of China (GJJ170925, GJJ180864).

References

1. Binggeli O, Neyen C, Poidevin M et al (2014) Prophenoloxidase activation is required for survival to microbial infections in *Drosophila*. PLoS Pathog 10(5):e1004067
2. Dudzic JP, Kondo S, Ueda R et al (2015) *Drosophila* innate immunity: regional and functional specialization of prophenoloxidases. BMC Biol 13:81
3. González-Santoyo I, Córdoba-Aguilar A (2012) Phenoloxidase: a key component of the insect immune system. Entomol Exp Appl 142(1):1–16
4. Cerenius L, Lee BL, Soderhall K (2008) The proPO-system: pros and cons for its role in invertebrate immunity. Trends Immunol 29(6):263–271
5. An C, Kanost MR (2010) *Manduca sexta* serpin-5 regulates prophenoloxidase activation and the Toll signaling pathway by inhibiting hemolymph proteinase HP6. Insect Biochem Mol Biol 40(9):683–689
6. Nakhleh J, El Moussawi L, Osta MA (2017) The melanization response in insect immunity. In: Ligoxygakis P (ed) Insect immunity. Advances in insect physiology, vol 52. Academic, London, pp 83–109
7. Lu A, Zhang Q, Zhang J et al (2014) Insect prophenoloxidase: the view beyond immunity. Front Physiol 5:252
8. Li YC, Wang Y, Jiang HB et al (2009) Crystal structure of *Manduca sexta* prophenoloxidase provides insights into the mechanism of type 3 copper enzymes. Proc Natl Acad Sci U S A 106(40):17002–17006
9. Aguilera F, McDougall C, Degnan BM (2013) Origin, evolution and classification of type-3 copper proteins: lineage-specific gene expansions and losses across the Metazoa. BMC Evol Biol 13:96
10. Yang B, Lu A, Peng Q et al (2013) Activity of fusion prophenoloxidase-GFP and its potential applications for innate immunity study. PLoS One 8(5):e64106
11. Masuda T, Momoji K, Hirata T et al (2014) The crystal structure of a crustacean prophenoloxidase provides a clue to understanding the functionality of the type 3 copper proteins. FEBS J 281(11):2659–2673
12. Wu K, Zhang J, Zhang Q et al (2015) Plant phenolics are detoxified by prophenoloxidase in the insect gut. Sci Rep 5:16823
13. Zhang J, Huang W, Yuan C et al (2017) Prophenoloxidase-mediated *ex vivo* immunity to delay fungal infection after insect ecdysis. Front Immunol 8:1445
14. Shao Q, Yang B, Xu Q et al (2012) Hindgut innate immunity and regulation of fecal microbiota through melanization in insects. J Biol Chem 287(17):14270–14279
15. Zhang J, Lu A, Kong L et al (2014) Functional analysis of insect molting fluid proteins on the protection and regulation of ecdysis. J Biol Chem 289(52):35891–35906
16. Liu D, Wang L, Yang L et al (2015) Serpin-15 from *Bombyx mori* inhibits prophenoloxidase activation and expression of antimicrobial peptides. Dev Comp Immunol 51(1):22–28
17. Lu A, Li X, Hillyer JF et al (2014) Recombinant *Drosophila* prophenoloxidase 1 is sequentially cleaved by alpha-chymotrypsin during *in vitro* activation. Biochimie 102:154–165
18. Wu K, Han F, Yuan Y et al (2018) Effect of the insect phenoloxidase on the metabolism of l-DOPA. Arch Insect Biochem Physiol 98(4):e21457
19. Liu F, Huang W, Wu K et al (2017) Exploiting innate immunity for biological pest control. In: Ligoxygakis P (ed) Insect immunity. Advances in insect physiology, vol 52. Academic, London, pp 199–230
20. Li X, Ma M, Liu F et al (2012) Properties of *Drosophila melanogaster* prophenoloxidases expressed in *Escherichia coli*. Dev Comp Immunol 36(4):648–656

21. Diao Y, Lu A, Yang B et al (2012) Existence of prophenoloxidase in wing discs: a source of plasma prophenoloxidase in the silkworm, *Bombyx mori*. PLoS One 7(7):e41416

22. Liu F, Chen Y, Yang B et al (2012) *Drosophila melanogaster* prophenoloxidases respond inconsistently to Cu^{2+} and have different activity *in vitro*. Dev Comp Immunol 36(3):619–628

Chapter 8

Basic Methods to Evaluate Humoral Immunity Processes in Lepidoptera Larvae

Maristella Mastore and Maurizio Francesco Brivio

Abstract

Even though insects lack an adaptive immune system, they can survive in environments with many potentially pathogenic invaders. Insects challenge infections by innate immunity defense mechanisms. Among them hemolymph humoral components cooperate to perform melanization (humoral encapsulation) and bacterial clearance. Investigating these two processes is fundamental to understand and check the insect physiological condition either normal or altered by infections or environmental changes. Many experimental protocols to investigate humoral defenses in insects are present in the literature, but discrepancies between them often exist. Such discrepancies are mostly due to the different biology of animal models and to the nature of experimental approach. Here we described less time-consuming and cheaper protocols used to test the activity of both constituent and inducible humoral components present in the hemolymph of insect larvae.

Key words Insect immunity, Hemolymph, Melanization, Bacterial clearance, proPO system, Phenoloxidase, Lysozyme, Antimicrobial peptides

1 Introduction

1.1 Insect Humoral Immunity

As a result of an infection or parasitization, insects react by triggering an innate immune response leading to the synthesis, and release in the hemolymph, of a pool of molecules (humoral factors) by the immunocompetent cells. The humoral factors function as immune molecules that cooperate in a synergistic process with the hemocytes populations to recognize and eliminate the invaders [1, 2]. Lysozyme, phenoloxidase, and antimicrobial peptides (AMPs) are components of the humoral response, whereas phagocytosis, encapsulation, and nodulation are processes carried out by insects hemocytes (Fig. 1). The methods described in this chapter allow to study some immune processes that take place in insect circulatory system (hemocoel), without requiring purification steps of molecular components. These methods can thus be used to analyze, by less time-consuming and less costly assays, the basal processes

Federica Sandrelli and Gianluca Tettamanti (eds.), *Immunity in Insects*, Springer Protocols Handbooks,
https://doi.org/10.1007/978-1-0716-0259-1_8,

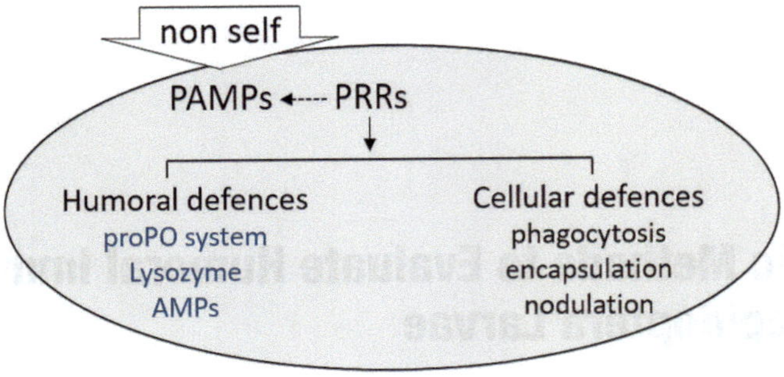

Fig. 1 Graphic summary of the main humoral and cellular immune defenses of insects triggered by the presence of foreign bodies (nonself)

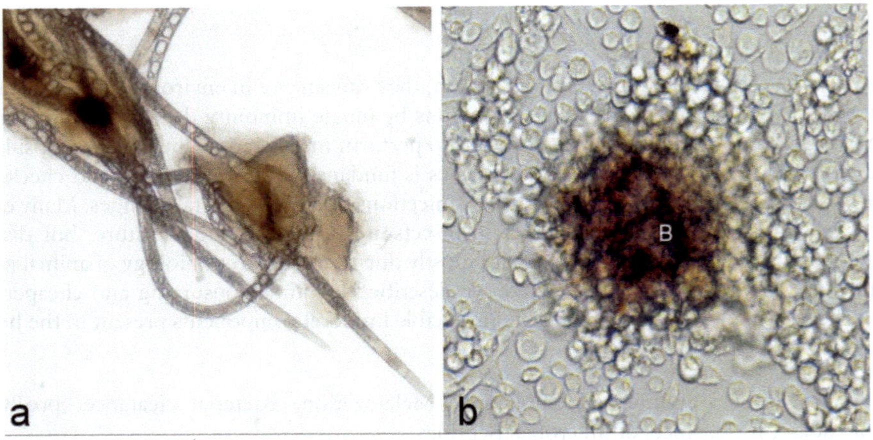

Fig. 2 Micrographs show melanization processes: brownish melanin compounds surround nematodes (left, **a**) and encase a synthetic microbead (right, **b**) encapsulated by the host hemocytes

occurring in the hemolymph and their possible modulation induced by microorganisms, parasites, and/or environmental stress. All the protocols described in this chapter can be applied to all species of Lepidoptera with the appropriate methodological variations, such as hemolymph sampling method, concentration of activators, bacterial load in experimental infections, etc.

1.2 Melanization and proPO System

The insect humoral encapsulation is based on melanization processes resulting from the activation of the hemolymph prophenoloxidase-phenoloxidase system (proPO system) which is responsible for the synthesis and deposition of melanin compounds around foreign bodies (Fig. 2a) and in the late stages of cellular encapsulation (Fig. 2b). The proPO system is a complex enzyme cascade in which the last active enzyme (phenoloxidase) can oxidize phenols into quinones that in turn autocatalyze into melanin [3–5].

Since the proPO system reacts quickly to the presence of foreign bodies, the modulation of its activity is an extremely useful parameter to assess the level of an infection or possible immune-depressive effects induced by the presence of foreign invaders.

The activity of the proPO system can be studied with different approaches. Since the effects of the activation of prophenoloxidase are evident, as shown in Fig. 2, the process of melanization in the humoral encapsulation can be monitored simply by optical microscopy techniques: the brownish melanic compounds are easily observable in samples of insect hemolymph after extraction.

Electrophoretic techniques can be used to ascertain the presence of the enzyme in the hemolymph. Denaturing electrophoresis (SDS-PAGE: in the presence of denaturants and reducing agents) allows to determine the molecular weight of separated compounds and thus to identify putative bands of phenoloxidase. Moreover, native-PAGE can be used to ascertain the presence of phenoloxidase in situ, monitoring its activity detected through the formation of dopachrome in the gel (Fig. 3) [4].

However, if the goal is a proper evaluation of variations in phenoloxidase activity in naïve or treated larvae, after the administration of drugs, infection with bacterial strains, parasitization with metazoan, and metabolic or environmental changes, a spectrophotometric analysis is strongly recommended [6]. Spectrophotometric analysis of the relative activity of the proPO system is based on the Absorbance produced by reaction intermediates (quinone products) which have a peak of absorption at $\lambda = 490$ nm (Fig. 4).

1.3 Antibacterial Humoral Defenses

1.3.1 Lysozyme Activity

Lysozyme is a ubiquitous component of the bacteriolytic armory of insects. It is normally present in the hemolymph, and after an infection it is strongly induced together with other AMPs. This is the case for most insects, with some exceptions such as *Drosophila*: the transcription levels of its lysozyme genes decrease after bacterial injections into the hemocoel [7, 8].

Lysozyme can damage peptidoglycans, a specific component of bacterial cell wall. Peptidoglycan is composed of repeating amino sugars, *N*-acetylglucosamine (NAG) and *N*-acetylmuramic acid (NAM), cross-linked by peptide bridges. Lysozyme acts by hydrolyzing the bond between NAG and NAM. The action of the enzyme results in an increase in cell permeability that leads to the death of bacteria. Because of its specificity for peptidoglycans, lysozyme is particularly efficacious against Gram-positive bacteria.

Lysozyme activity assays are essentially based on two principles: (a) assay methods based on the protein itself, such as electrophoretic, chromatographic, immunoenzymatic, and spectro-photofluorometric techniques, and (b) assay methods relying on the lytic activity of lysozyme against the cell wall of the bacterium *Micrococcus lysodeikticus*, used as a substrate [9, 10].

Considering its relatively easy and low costs, in this chapter we described the turbidimetric method, as it is particularly suitable for comparing the relative activity of lysozyme in samples of hemolymph

Fig. 3 Non-denaturing electrophoresis (native-PAGE) of a sample of cell-free hemolymph. The lane at the left was stained by Coomassie blue; lane at the right shows one band (red arrow) positive for phenoloxidase activity after enzymatic staining with L-Dopa as substrate

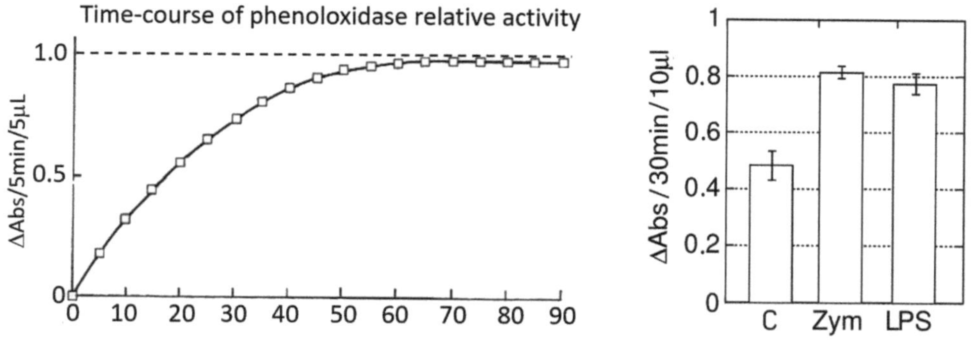

Fig. 4 Left: Phenoloxidase relative activity was recorded every 5 min and plotted on a graph; the curve shows the increase in Absorbance over time. The conversion rate of L-Dopa to dopachrome is initially high and gradually decreases until it reaches a plateau. Right: The graph shows the variation of basal phenoloxidase activity in naïve larvae (C) and in larvae injected with proPO system activators (PAMPs) such as zymosan (Zym) and Gram-negative bacteria lipopolysaccharides (LPS). The assays were carried out with hemolymph obtained 30 min after PAMPs injection, and ΔAbs was recorded 30 min after the start of the reaction

from larvae. The method is based on the turbidity of a *M. lysodeikticus* suspension and on its decrease due to the lytic action of the lysozyme. Changes in turbidity can be recorded as downward variation of Absorbance (Fig. 5) by a spectrophotometer or a microplate Absorbance reader.

1.3.2 Antimicrobial Peptides (AMPs)

Insect AMPs are mainly synthesized in fat body tissues and released into the hemolymph during a systemic response against pathogens (Gram-negative, Gram-positive bacteria and fungi). The sensing of foreign bodies culminates in the synthesis ex novo of AMPs; thus, these are considered inducible factors [11]. Their occurrence is consequent to the activation of immune genes mediated by evolutionarily conserved Toll/Imd pathways present in both vertebrates and invertebrates [12]. AMPs are strong cationic, heat-stable, and amphipathic small molecules that have a variable amino acid composition, length, and structure [13]; in general, most AMPs cause serious damages to the wall leading bacteria to death (Fig. 6).

Investigations of the AMPs occurrence and activity can be addressed in different ways by means of electrophoretic methods or microbiological assays, respectively. The presence in the hemolymph and the preliminary characterization of AMPs can be carried out by a modification (Tricine-PAGE) of the conventional electrophoretic analysis (SDS-PAGE) [14–16]. The difference in proteins separation obtained by SDS-PAGE (Fig. 7, left) or Tricine-PAGE are based on the different pK values of the glycine and tricine ions, affecting their electrophoretic mobilities and thus resulting in a different mobility of the proteins. Tricine-PAGE produces an improved resolution in the separation of proteins with a molecular weight below 30 kDa (Fig. 7, right).

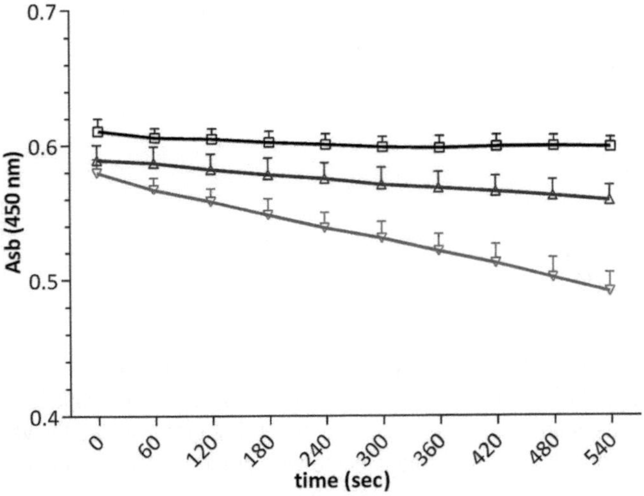

Fig. 5 The graph shows the increase of lysozyme activity in the hemolymph of insect larvae after bacterial infection (blue line); Absorbance decreases faster with respect to naïve larvae control sample (red line). Black line shows the Absorbance of *M. lysodeikticus* suspension (as a control)

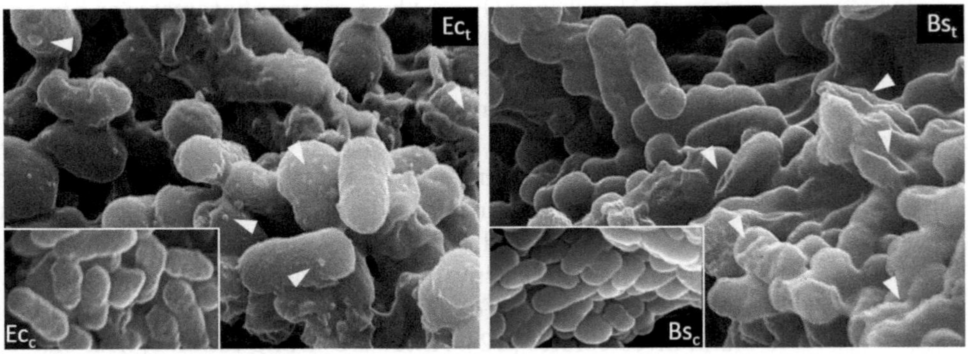

Fig. 6 Scanning electron microscopy (SEM) micrographs show the effects of AMPs treatments on Gram-negative (*Escherichia coli*, Ec; image at the left) and Gram-positive (*Bacillus subtilis*, Bs; image at the right) bacteria. Arrowheads at the left (Ec_t) indicate blebs on the surface of the bacterial wall and at the right (Bs_t) collapsed bacterial cells. Controls (untreated) are shown in insets

Concerning the activity of the AMPs present in the hemolymph, the most common methods are based on microbiological tests that allow to evaluate the lethal effects of AMPs on different bacterial strains [16]. The main microbiological methods come from the refinement and/or modification of classical methods such as radial diffusion, microbroth dilution assay, and gel overlay assay [17]. Evaluation of colony-forming units (CFU) by means of serial dilution plating on a nutrient medium is one of the most widely accepted procedures to monitor bacteria growth and viability (Fig. 8). In this chapter we described a modification of the microbroth dilution assay (single plate-serial dilution spotting),

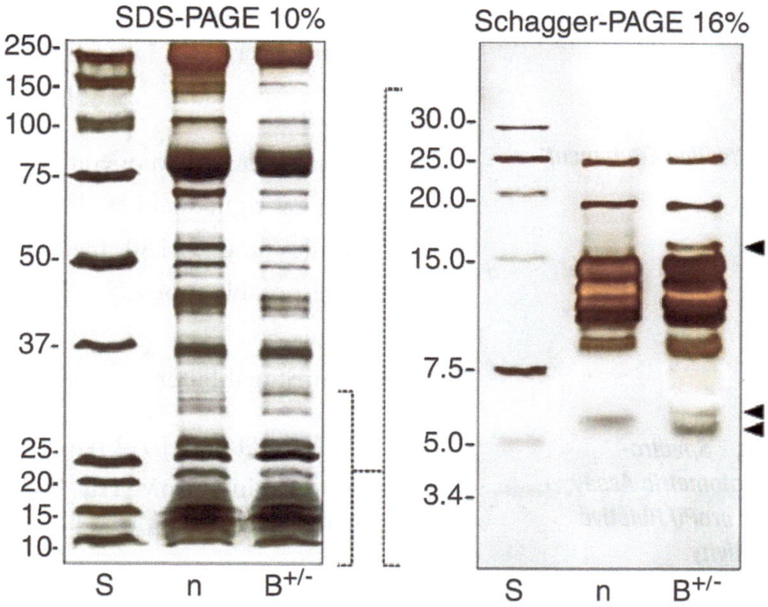

Fig. 7 Difference in resolution of low molecular weights proteins (<30 kDa) after separation by 10% SDS-PAGE (left) or by 16% Tricine-PAGE (right). Samples are from insect hemolymph before (n) and after bacterial infection ($B^{+/-}$). The hemolymph samples analyzed by Schägger-PAGE (Tricine) have been previously fractioned (<30 kDa) by ultrafiltration

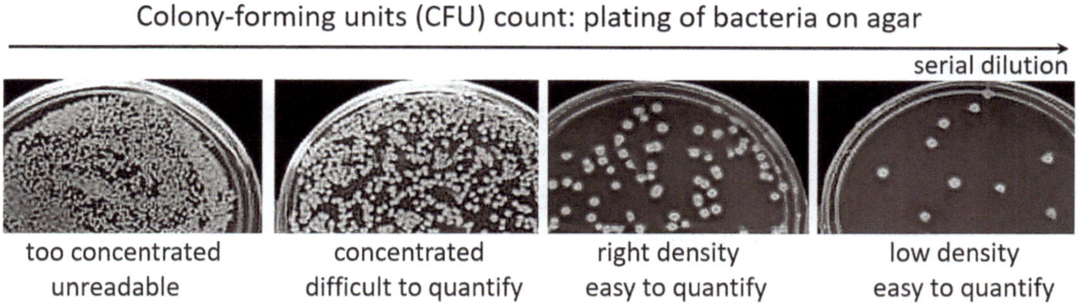

Fig. 8 When bacteria are plated on agar, if properly diluted, they form a number of easily quantifiable colonies (images at the right). The number of colonies (CFU) reflects the concentration of cells in the culture; thus, if bacteria are treated with antimicrobials, the reduction of live cells is easily assessed

carried out with microdilutions and drops spotted on vertically positioned plates, named track-dilution method. This is a simple procedure to quantify viable bacteria on agar plates.

2 Materials

2.1 Insect Models

Larvae of Lepidoptera are reared on proper diet and under controlled conditions (temperature, humidity). Based on the planned study, different instars from healthy larvae can be used in all the assays. Diet and rearing conditions could be different for different species of Lepidoptera.

2.2 Hemolymph Collection Equipment

1. Ethanol 70%.
2. Ice or cooling blocks to operate in cold condition (4 °C).
3. Hamilton gas tight syringes (Hamilton, Reno, NE, USA).
4. Thin needles (cod. 90130, Hamilton, Reno, NE, USA).
5. 1.5-mL disposable tubes.
6. Centrifuge.
7. Ultra Centrifugal Filters (cutoff 50–30–10 kDa).

2.3 Spectrophotometric Assay for proPO Relative Activity

1. 100 mM Tris–HCl pH 7.2 (stock solution).
2. L-Dopa 8 mM in 10 mM Tris–HCl pH 7.2.
3. Double beam spectrophotometer.
4. 1.4-mL disposable microcuvettes.
5. As activators (if needed). Zymosan: 2 mg of zymosan, β-glucans from *Saccharomyces cerevisiae* dissolved in 1 mL of sterile 10 mM Tris–HCl pH 7.2, homogenized for 15 min in ultrasound bath, and centrifuged at 14,000 × *g* for 5 min to recover the supernatant; or LPS: 5 mg of LPS from *Escherichia coli* O111:B4 dissolved in 1 mL of 10 mM Tris–HCl pH 7.2.

2.4 In Situ Identification of Phenoloxidase by PAGE

1. Vertical electrophoresis cell.
2. Power supply.
3. One-dimensional polyacrylamide gel slabs.
4. 15- and 50-mL disposable tubes.
8. Ultrapure water.
5. Filter paper.
6. Thermostatic bath or thermodry bath.

2.4.1 Phenoloxidase Activity Detection by Native-PAGE

1. Native gels. 10% resolving gel: 8.45 mL of ultrapure water, 6.6 mL of 30% acrylamide/bis-acrylamide stock (29.2:0.8 ratio), 5.0 mL of 0.5 M Tris–HCl pH 8.8, 100.0 μL of 10% ammonium persulfate (APS), and 10.0 μL *N,N,N′,N′*-tetramethyl-ethylene-diamine (TEMED). 4% stacking gel: 4.5 mL of ultrapure water, 1.2 mL of 30% acrylamide/bis-acrylamide stock, 1.9 mL of 0.5 M Tris–HCl pH 6.8, 50.0 μL of 10% APS, and 5.0 μL TEMED.
2. 10× running buffer: dissolve in deionized water 30.0 g Tris-base (MW 121.14 g/mol), 144.0 g glycine (MW 75.07 g/mol), pH 8.8, 1-L final volume.

3. 2× native sample loading buffer: 62.5 mM Tris–HCl, pH 6.8, 40% (v/v) glycerol, 0.01% (w/v) bromophenol blue.
4. Phenoloxidase staining: Equilibrating buffer (20 mM Bis-Tris–HCl, pH 6.5), activating solution (0.5 mg/mL chymotrypsin in 20 mM Bis-Tris–HCl, pH 6.5), and development solution (1 mg/mL L-Dopa in 20 mM Bis-Tris–HCl, pH 6.5).

2.4.2 Phenoloxidase Detection by SDS-PAGE

1. Denaturing gels. 10% resolving gel: 8.3 mL of ultrapure water, 6.6 mL of 30% acrylamide/bis-acrylamide stock solution, 5.0 mL of 1.5 M Tris–HCl, 0.4% SDS, pH 8.8, 100.0 μL of 10% APS, and 10.0 μL TEMED. 4% stacking gel: 4.5 mL of ultrapure water, 1.2 mL of 30% acrylamide/bis-acrylamide stock solution, 1.9 mL of 0.5 M Tris–HCl, 0.4% SDS, pH 6.8, 50.0 μL of 10% APS, and 5.0 μL of TEMED.
2. 10× running buffer: dissolve in water 30.0 g Tris-base (MW 121.14 g/mol), 144.0 g glycine (MW 75.07 g/mol), 10.0 g SDS (MW 288.372 g/mol), pH 8.8, 1-L final volume.
3. 2× SDS sample loading buffer: 1.1 mL of distilled water, 2.4 mL of 0.5 M Tris–HCl pH 6.8, 4 mL of 10% (w/v) SDS, 20% (v/v) glycerol, 0.1% (w/v) bromophenol blue, 2% (v/v) 2-mercapto-ethanol (*see* **Note 1**).
4. Coomassie Staining. Staining solution (0.1% Coomassie Brilliant Blue R-250, methanol/glacial acetic acid/distilled water) (5:1:5); destaining solution: methanol/glacial acetic acid/distilled water (5:1:5); storage solution (5% glacial acetic acid).
5. Silver staining solution: Fixative solution (50% methanol, 12% acidic acid, 3.7 mL of 37% formaldehyde to prepare 100 mL of solution); washing solution (50% ethanol); pretreatment solution (sodium thiosulfate 20 mg/100 mL); ultrapure water; silver nitrate solution (200 mg silver nitrate, 75 μL of 37% formaldehyde/100 mL); developing solution (6 g sodium carbonate, 0.2 mg of sodium thiosulfate, 50 μL of 37% formaldehyde/100 mL); stop solution (50% methanol, 12% acidic acid); fixing solution (50% methanol); storage solution (deionized water) (*see* **Note 2**).

2.5 Antimicrobial Identification and Activity Assays

2.5.1 Lysozyme Activity: Turbidity Assay

1. Disposable conical tube (50 mL).
2. Flat-bottomed 96-well plate.
3. Microplate Absorbance reader.
4. 100 mM phosphate buffer, pH 6.8.
5. *Micrococcus lysodeikticus* ATCC No. 4698: 0.45 mg/mL of *M. lysodeikticus* in phosphate buffer.

2.5.2 AMPs Identification by Tricine-PAGE

1. Tricine-PAGE gels. 16% resolving gel: 6.66 mL of 48.2% acrylamide/1.8% bis-acrylamide stock solution, 6.66 mL of gel buffer (3 M Tris–HCl, 0.3% SDS, pH 8.45), 2.11 mL of glycerol, 4.57 mL of ultrapure water, 100.0 μL of 10% APS, and 10.0 μL TEMED. 4% stacking gel: 0.5 mL of acrylamide/bis-acrylamide stock solution, 1.55 mL of gel buffer, 4.2 mL of ultrapure water, 100.0 μL of 10% APS, and 10.0 μL TEMED.
2. Running buffers. Anodic buffer: 200 mM Tris-base, pH 8.9. Cathodic buffer: 100 mM Tris-base, 100 mM tricine, 0.1% SDS, pH 8.26.
3. Tricine sample loading buffer: 4 mL of ultrapure water, 2.0 mL of 0.5 M Tris–HCl, pH 6.8, 2.4 mL of glycerol, 1 mL of 10% (w/v) SDS, 0.2 mL of β-mercaptoethanol, 0.4 mL of 0.5% (w/v) Coomassie G-250.
4. To stain the gel with Coomassie or Silver staining methods, *see* Subheading 2.4.2, **item 4** or **5**.

2.5.3 AMPs Activity: CFU Count by Track-Dilution Method

1. Bunsen burner.
2. Sterile loops (wire or plastic).
3. Cell spreader (glass or plastic).
4. Pipettes (calibrated/dropping; glass/plastic).
5. Bottles (wide neck).
6. Erlenmeyer flasks.
7. Disposable conical tubes (15 mL).
8. Disposable Petri dishes or disposable square dishes.
9. Autoclave.
10. Benchtop Shaking Incubator.
11. Thermostatic bath.
12. Hamilton gas tight syringes (Hamilton, Reno, NE, USA).
13. Thin needles (Hamilton, Reno, NE, USA).
14. Parafilm.
15. Flat-bottomed 96-well plate.
16. Microplate Absorbance Reader.
17. Personal protective equipment.
18. *Escherichia coli* strain K12 (ATCC 10798), *Bacillus subtilis* (ATCC 6051), or other bacterial strains.
19. 1 M phosphate Buffer (PB) stock solutions: 136 g of KH_2PO_4 (MW 136.09) to 1 L of distilled water; 174.2 g of K_2HPO_4 (MW 174.18) to 1 L of distilled water. To obtain a 1× PB, at pH 7.2, mix 71.7 mL of K_2HPO_4 and 28.3 mL of KH_2PO_4, and dilute to 1 L.

20. Growth media. Liquid Luria Broth (LB): dissolve 20 g of LB (1% tryptone, 0.5% yeast extract, 0.5% NaCl) in 1 L of distilled water, shake to dissolve the solute, and sterilize by autoclave at 120 °C, for 15 min.
21. Agar plates: dissolve 20 g of LB and 15 g of agar in 1 L of distilled water, shake, and finally sterilize by autoclave at 120 °C, for 15 min (*see* **Note 3**).

3 Methods

3.1 Insect Hemolymph Collection

1. Surface-sterilize larvae with 70% ethanol (*see* **Note 4**).
2. Flush out the hemolymph in an ice-cold tube, to avoid undesired activation of proteases, by puncturing the larva with a fine needle (*see* **Note 5**).
3. Centrifuge whole hemolymph samples at 200 × *g*, for 10 min, at 4 °C, to remove cells and tissue debris.
4. Recover the supernatants (cell-free fraction, CFF).
5. (a) To analyze proPO system activity, CFF should be immediately assayed. (b) To assay lysozyme activity, addition of few crystals of *N*-phenylthiourea (PTU) is required to avoid the activation of prophenoloxidase. (c) To detect the activity and identify AMPs, the hemolymph (CFF plus PTU) can be further processed to separate samples in a defined molecular weights range, by centrifugal filter units with low molecular weight cutoff. CFF is centrifugated in the filter unit with a swinging bucket rotor, at 4000 × *g*, for 1 h, at 15 °C.

3.2 Spectrophotometric Assay of Phenoloxidase Relative (Basal) Activity

1. Prepare 8 mM of L-Dopa in 10 mM of Tris–HCl, pH 7.2 (*see* **Note 6**).
2. Set two blanks adding 1 mL of L-Dopa buffer to spectrophotometric cuvettes.
3. Set samples cuvettes adding 1 mL of L-Dopa buffer, and, when ready, add a variable volume (5–20 μL) of CFF to the L-Dopa, finally mix, and start the record (*see* **Note 7**).
4. Read immediately the increase in Absorbance (Abs), due to the dopachrome formation, at 490 nm by the spectrophotometer, and record the ΔAbs_{490nm}, at 20 °C, at specified time intervals (1–5–10 min), for a defined period, not exceeding 1.0 Abs (*see* **Note 8**).

3.2.1 Spectrophotometric Assay of Phenoloxidase Relative Activity: Modulation of the proPO System by PAMPs

The proPO system can be activated, over its basal level, by the presence of elicitors (PAMPs) or microorganisms.

If an in vivo assay is required:

1. Inject into larvae hemocoel an amount of the elicitor.

2. Wait a variable time (indicatively from 10 to 60 min) and collect the hemolymph as described at Subheading 3.1.
3. Proceed as described at Subheading 3.2, **steps 2–4**.

If an in vitro assay is required:

1. Add an amount of elicitor in the mix L-Dopa buffer (1 mL) plus CFF, mix and incubate for 15–20 min at room temperature, and then start the record. Alternatively the Abs recording can be started immediately after addition of the elicitor.
2. Proceed as described at Subheading 3.2, **step 4**.

All samples should be assayed in triplicate. Plot the ΔAbs_{490nm}/time/CFF volume.

3.3 In Situ Identification of Phenoloxidase or AMPs: Electrophoretic Techniques

3.3.1 Sample Preparation

1. Prepare a suitable amount of sample loading buffer (for SDS, native- or Tricine-PAGE).
2. Collect and process the hemolymph as in Subheading 3.1, **steps 1–4**.
3. Determine protein concentration in the CFF by Bradford method [18].
4. Add an amount of CFF to the loading buffer 1× (indicatively 1:20, based on the total proteins in CFF).
5. Centrifuge samples at 10,000 × *g* for 30 s to remove insoluble particles.
6. Samples are ready to load for native-PAGE.
7. For SDS- or Tricine-PAGE, denaturate samples for 5–10 min at 100 °C in a thermostatic bath.
8. Centrifuge samples at 10,000 × *g* for 30 s to remove insoluble material.
9. Use immediately or store all samples at −20 °C.

3.3.2 Electrophoresis (SDS-, Tricine-, and Native-PAGE)

1. Assemble the glass plate sandwich (following the manufacturer instruction).
2. Prepare the resolving gel solution in a tube (50 mL) as described in Subheading 2.4.
3. Add APS and mix, and then add TEMED and mix carefully (avoid formation of bubbles) (*see* **Note 9**).
4. Pour the gel solution between the glass plates with a Pasteur pipette, and leave about 1/4 of the space free for the stacking gel (*see* **Note 10**).
5. Cover the top of the resolving gel with a layer (1–2 cm) of ultrapure water, and wait until the gel polymerizes (20–30 min).
6. When polymerization is complete, a sharp line appears between gel and water.

7. Remove water, then clean, and dry water residues between the plates with filter paper sheets.
8. Prepare the stacking gel solution in a tube (15 mL) as described in Subheading 2.4.
9. Add APS and TEMED, and mix carefully to avoid formation of bubbles (*see* **Note 9**).
10. Before polymerization, to make wells insert a comb in the stacking gel; carefully avoid bubbles and wait for polymerization for at least 20–30 min.
11. When ready to load, remove the comb, wash twice the wells with running buffer, and finally refill the wells.
12. Load with a micropipette samples and protein marker inside the wells.
13. Put the gel into the electrophoresis tank, fill the tank (bottom and top reservoirs) with running buffer, and make sure that samples in the wells and the wire electrode of the upper chamber are immersed in the buffer (*see* **Note 11**).
14. Set an appropriate voltage or current, depending on how many gels you run. Start the electrophoresis at low voltage (50–70 V), and then increase the power when the dye front reaches the running gel (*see* **Note 12**).
15. Stop the electrophoresis when the dye front reaches the bottom of the gel. Disassemble the gel sandwich and proceed with gel staining.

3.3.3 Gel Staining

Staining for Phenoloxidase Activity (Native-PAGE)

1. Soak the gel in the equilibrating buffer for 15 min.
2. Remove the equilibrating buffer and soak the gel in the activating solution, for 15 min at 20 °C.
3. Incubate the gel with development solution (L-Dopa) at 20 °C, in the dark, and wait for band(s) to appear.
4. Wash and store the gel in deionized water.

Staining for SDS- and Tricine-PAGE (Denaturing-PAGE)

Denaturing gels can be stained by two main protocols:

1. Coomassie staining: wash the gel briefly with deionized water; soak the gel in staining solution for 20–60 min (depending from the gel thickness) at room temperature on a shaker; remove the staining solution and add the destaining solution until blue background disappears and protein bands are visible; all steps are carried out under shaking. Store the gel in storage solution.
2. Silver staining: fix the gel in fixative solution for 1 h at room temperature; remove the fixative and wash by washing solution (three times, 20 min each one). Remove the washing solution and add the pretreatment solution (1–2 min); wash

with ultrapure water (3 0 s each one). Incubate in silver nitrate solution (20 min, in the dark). Remove the silver nitrate and incubate in developing solution, wait for bands to appear, stop the reaction, and store in ultrapure water. All steps are carried out under shaking.

3.4 Hemolymph Antimicrobial Activity

3.4.1 Assay of Lysozyme Activity

1. Dissolve 0.45 mg/mL of dried *M. lysodeikticus* cells in 0.1 M phosphate buffer, pH 6.8 (this cells concentration gives an Abs of about 0.6–0.7 at λ = 450 nm).
2. Add 190 μL of *M. lysodeikticus* cell suspension to wells of a flat-bottomed 96-well plate.
3. Add an amount of hemolymph samples (e.g., 10 μL) from untreated or treated (pre-injected with PAMPs or microorganisms) larvae; mix the solution.
4. For the blank, add 190 μL of *M. lysodeikticus* cell suspension into well, and add a volume of phosphate buffer matching that of the added hemolymph (*see* **Note 13**).
5. Read immediately the Abs of samples, and record the lysozyme activity at 30-s intervals, for a short time period (2–10 min) by a microplate Absorbance reader (Fig. 9) (*see* **Note 14**).
6. For each sample at least three repetitions of Abs recording are performed, and then the average value was considered (*see* **Note 15**).

3.4.2 Bacterial Culture

1. Prepare LB medium in a sterile flask.
2. Autoclave the broth and cool to room temperature.
3. Inoculate bacteria (*E. coli* or *B. subtilis*) with sterile loop from bacteria frozen glycerol stock solution into 50 mL of LB.

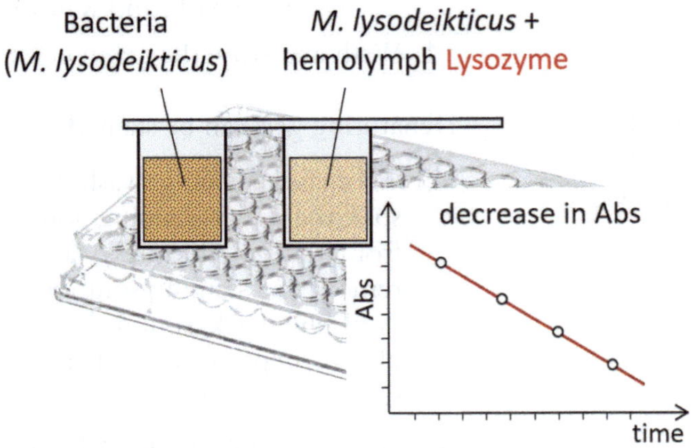

Fig. 9 Decrease in turbidity in bacteria samples after the addition of hemolymph lysozyme; the activity of the enzyme can be quantified spectrophotometrically by Absorbance variation

4. Grow the culture overnight under shaking at 37 °C, in a dark room.
5. Determine the optical density (O.D.) of the culture broth by spectrophotometry at $\lambda = 600$ nm.
6. To standardize the assay, prepare a blank with 1 mL of sterile nutrient broth.
7. For antimicrobial activity assays, use O.D. = 0.6 (*see* **Note 16**).

3.4.3 Preparation of Bacteria and Larvae Immunization

1. Put 1 mL of bacterial culture (10^9 CFU) in tube, pellet by centrifugation, at 1700 × *g*, for 10 min, at 20 °C.
2. Remove supernatant and wash the bacteria pellet three times with phosphate buffer.
3. Put the bacteria tube in a thermostatic bath and kill bacteria by heating at 65 °C for 2 h.
4. Wash the bacterial pellet with phosphate buffer and prepare a mixture 1:1 of *E. coli* and *B. subtilis.*
5. To inject the larvae, prepare a bacteria suspension with a final concentration of 10^5 CFU/mL.
6. Surface-sterilize a number of larvae with ethanol, and inject a number of bacteria (10^2–10^3 CFU) by a microsyringe (*see* **Note 17**).
7. For control, inject the same amount of PBS buffer into larvae, or prick the larvae with a sterile needle.
8. Put larvae for 30 s on ice to reduce the loss of hemolymph.
9. Incubate larvae 12–48 h under rearing condition.
10. Surface-sterilize the larvae with ethanol, and then bleed by puncturing the larvae by a sterile needle as described at Subheading 3.1, **steps 1–4**.
11. If needed, process the hemolymph by ultrafilter devices to obtain low molecular weight (30 or 10 kDa) fractions, as described at Subheading 3.1, **step 5**.
12. Quantify total proteins content by Bradford protein assay [18].

3.4.4 Antimicrobial Activity in Hemolymph Samples

1. Prepare bacterial culture as described in Subheading 3.4.2.
2. Determinate the O.D. of bacterial culture.
3. Prepare serial dilutions of the bacterial culture.
4. Put in a sterile tube 9.9 mL of LB broth and add 100 μL of bacterial culture, dilution 1:100, to reach 10^7 CFU/mL.
5. Put in a second tube 9 mL of LB broth, and add 1 mL of 10^7 CFU/mL bacterial culture, dilution 1:10, to obtain 10^6 CFU/mL (CFU at time zero).
6. Set samples. Put 180 μL of 10^6 CFU/mL bacterial culture in a 0.5-mL tube, and add 20 μL (per μg of proteins) of LMW

hemolymph (or whole hemolymph); vortex 30 s. Samples are from naïve or immunized (treated) larvae.

7. Set the blank. Put 180 μL of 10^6 CFU/mL bacterial culture in a 0.5-mL tube, and add 20 μL of phosphate buffer; vortex 30 s.
8. Set the sterile control, 200 μL of LB broth.
9. Incubate all samples in a shaking incubator for 3 h at 37 °C.

3.4.5 Evaluation of Bacteria CFU: Track-Dilution Method

1. Prepare microdilutions in a 96-well microplate.
2. In one well of first column, add 100 μL of bacterial culture (10^6 CFU/mL), or 100 μL of bacteria + PB, preincubated for 3 h, or 100 μL of bacteria + AMPs (hemolymph), preincubated for 3 h. Dispense 90 μL of phosphate buffer in each well, from column 2 to column 8, for each row.
3. Proceed with the serial dilutions following a procedure similar to that illustrated in Fig. 10.
4. Prepare two agar round plates for each sample (or use a squared plate).
5. Pick up 10 μL from each well and put the drop onto the agar plate, and then place the plate vertically to slide the drop. Seal each plate with parafilm.
6. Incubate the plates for 24 h at 37 °C.
7. Count the number of colonies (CFU) in each plate.
8. Calculate the effective dilution: CFU × dilution factor = CFU/mL.

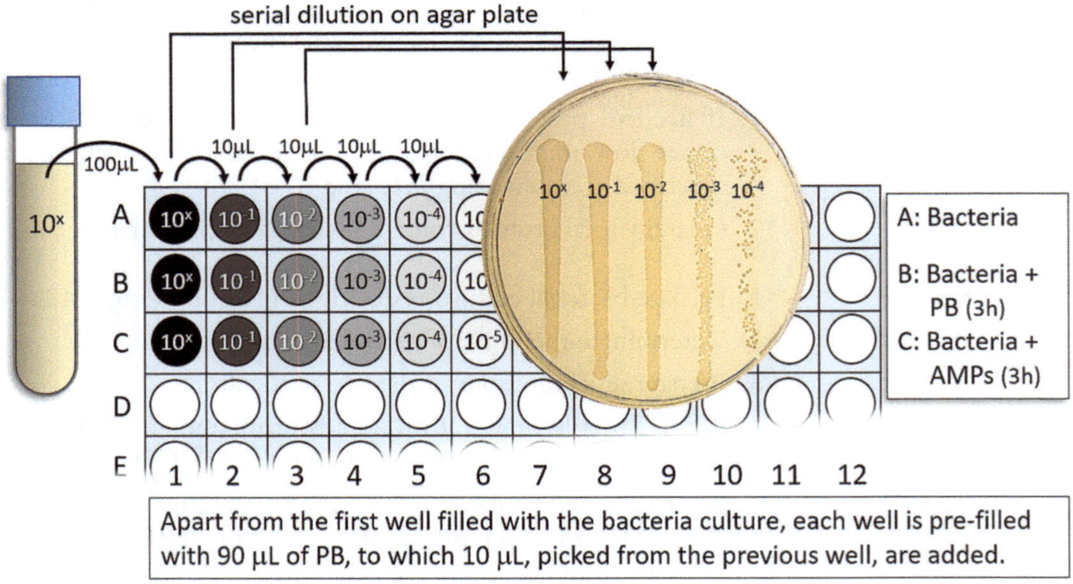

Fig. 10 The scheme summarizes the serial dilutions in 96-well plate followed by the plating on agar Petri dishes by the track-dilution method

The antibacterial activity in hemolymph samples is intended as the percentage of bacterial survival compared with the control.

4 Notes

1. 2-Mercaptoethanol is not stable in solution; thus it should be added to the denaturing buffers immediately before use.
2. To prepare the Silver staining solution, use only ultrapure water.
3. Use only sterile materials; filter not autoclavable buffers and solutions on 0.22 μm filter devices.
4. After the sterilization by ethanol, wash the larvae with sterile water, and then dry the body with soft paper sheets.
5. Body region, depth, and angle of the prick may vary depending on instar, size, and species of the insect.
6. L-Dopa buffer must be prepared freshly before the assay and kept in the dark.
7. Different volumes of pure or diluted hemolymph can be tested, according to insect species, larval instar, and physiological conditions.
8. If the Absorbance of the sample exceed 1.0 units, reduce the amount of CFF in the reaction L-Dopa buffer.
9. After the addition of APS and TEMED polymerization begins, thus all subsequent actions must be performed promptly.
10. If a Protean (Bio-Rad) cell is used, volumes of the gel solutions described in Subheading 2.4 are already calibrated for the correct dimensions of the glass plates.
11. For Tricine-PAGE, running buffers for the upper and lower chambers of the apparatus are different: cathodic buffer in the upper and anodic buffer in the lower chamber.
12. High voltage produces heat; do not exceed (unless you use a cooling system) 120–150 constant voltage.
13. Volumes of hemolymph and bacteria suspension can be modified, depending on the insect species under study (concentration and activity level of the enzyme in the hemolymph of different biological models).
14. Duration and time intervals of the assay may vary according to the enzyme activity in the biological models studied.
15. One unit of lysozyme activity (units/min/mL) is defined as the change of 0.001 units of Absorbance of a suspension of the Gram-positive bacteria *M. lysodeikticus* (ΔAbs450/min).
16. Maintain sterile conditions throughout all subsequent working steps, clean surfaces, and use a Bunsen burner to manipulate and transfer bacteria.

17. The number of CFU to be injected can be different, depending on the insect species, larvae instar, bacteria strain, or the immune process under study.

References

1. Hoffmann JA, Reichhart JM, Hetru C (1996) Innate immunity in higher insects. Curr Opin Immunol 8:8–13
2. Strand MR (2008) The insect cellular immune response. Insect Sci 15:1–14
3. Cerenius L, Lee BL, Söderhäll K (2008) The proPO-system: pros and cons for its role in invertebrate immunity. Trends Immunol 29:263–271
4. Brivio MF, Mazzei C, Scarì G (1996) proPO system of Allogamus auricollis (Insecta): effects of various compounds on phenoloxidase activity. Comp Biochem Physiol B 113:281–287
5. Cerenius L, Söderhäll K (2004) The prophenoloxidase-activating system in invertebrates. Immunol Rev 198:116–126
6. Brivio MF, Pagani M, Restelli S (2002) Immune suppression of Galleria mellonella (Insecta, Lepidoptera) humoral defenses induced by Steinernema feltiae (Nematoda, Rhabditida): involvement of the parasite cuticle. Exp Parasitol 101:149–156
7. Kylsten P, Kimbrell DA, Daffre S et al (1992) The lysozyme locus in Drosophila melanogaster: different genes are expressed in midgut and salivary glands. Mol Gen Genet 232(3):335–343
8. Hultmark D (1996) Insect lysozymes. EXS 75:87–102
9. Mörsky P (1983) Turbidimetric determination of lysozyme with Micrococcus lysodeikticus cells: re-examination of reaction conditions. Anal Biochem 128:77–85
10. Helal R, Melzig MF (2008) Determination of lysozyme activity by a fluorescence technique in comparison with the classical turbidity assay. Pharmazie 63:415–419
11. Bulet P, Hetru C, Dimarcq JL et al (1999) Antimicrobial peptides in insects, structure and function. Dev Comp Immunol 23:329–344
12. Medzhitov R, Janeway CA Jr (2002) Decoding the patterns of self and nonself by the innate immune system. Science 296:298–300
13. Bulet P, Stöcklin R (2005) Insect antimicrobial peptides: structures, properties and gene regulation. Protein Pept Lett 12:3–11
14. Laemmli UK (1970) Cleavage of structural proteins during assembly of the head of bacteriophage T4. Nature 227:680–685
15. Schägger H, von Jagow G (1987) Tricine-sodium dodecyl sulfate polyacrylamide gel electrophoresis for the separation of proteins in the range from 1–100 kDalton. Anal Biochem 166:368–379
16. Mastore M, Binda Rossetti S, Giovannardi S et al (2015) Inducible factors with antimicrobial activity after immune challenge in the haemolymph of Red Palm Weevil (Insecta). Innate Immun 21(4):392–405
17. Thomas P, Sekhar AC, Upreti R et al (2015) Optimization of single plate-serial dilution spotting (SP-SDS) with sample anchoring as an assured method for bacterial and yeast cfu enumeration and single colony isolation from diverse samples. Biotechnol Rep 8:45–55
18. Bradford MM (1976) A rapid and sensitive method for the quantitation of microgram quantities of protein utilizing the principle of protein-dye binding. Anal Biochem 72:248–254

Part IV

Humoral Response: Antimicrobial Peptides as Alternatives to Conventional Antibiotics

Chapter 9

Methods for the In Vitro Examination of the Antibacterial and Cytotoxic Activities of Antimicrobial Peptides

Bruno Casciaro, Floriana Cappiello, Maria Rosa Loffredo, and Maria Luisa Mangoni

Abstract

Antimicrobial peptides (AMPs) represent an interesting class of molecules with expanding properties. Nature is the primary source of AMPs since they are produced by most living organisms including prokaryotes, plants, and animals. Thanks to their hundreds of thousands of species on earth, insects are one of the most abundant and varied resources of AMPs. Among these, many families have already been well characterized while new AMPs are continuously discovered. In this chapter, the main methods for the in vitro evaluation of the biological properties of AMPs are described. In particular, to examine the antimicrobial activity, the inhibition zone assay and the techniques for the determination of the minimal inhibitory concentration and the bactericidal concentration are reported in detail. For the evaluation of the possible cytotoxic effect toward mammalian cells, the hemolytic test and the colorimetric assay based on the reduction of 3-(4,5-dimethylthiazol-2yl)-2,5-diphenyltetrazolium bromide are also described.

Key words Antimicrobial peptides, Insect immunity, Inhibition zone assay, Minimal inhibitory concentration, Bactericidal concentration, Hemolysis, Colorimetric MTT assay

1 Introduction

Antimicrobial peptides (AMPs) are multifunctional components of the innate immune system and act as a first line of defense against pathogens [1]. They are produced by most living organisms including prokaryotes, fishes, reptiles, amphibians, plants, mammals, and insects [2]. Insects represent the largest and most diverse group of organisms on earth with several million species in total. The amount and the types of AMPs produced by every species are linked to the nature of the environmental threats they faced during evolution. Considering the amount of existing insect species and their versatility in interacting with the surrounding environment, they represent one of the richest natural sources of bioactive

Bruno Casciaro and Floriana Cappiello contributed equally to this work.

Federica Sandrelli and Gianluca Tettamanti (eds.), *Immunity in Insects*, Springer Protocols Handbooks, https://doi.org/10.1007/978-1-0716-0259-1_9, © Springer Science+Business Media, LLC, part of Springer Nature 2020 compounds [3]. They are produced and secreted from cells and tissues of the innate immune system (e.g., hemocytes or fat body) and are involved in all defense-related processes such as direct killing of pathogens, modulation of the immune response, and endotoxin neutralization. Although pathogens are continuously exposed to these peptides, they can very rarely become resistant to this class of compounds. This is due to the mechanism of action of AMPs, which consists in the perturbation of the bacterial plasma membrane rather than a specific interaction with a cellular target that could mutate [4–6]. In the era of the antibiotic-resistance crisis in which we are living, the discovery of new antimicrobials is highly needed, and insect AMPs hold great promise [7].

Although several classes of AMPs derived from insects have already been well characterized (i.e., cecropins, drosocins, attacins, diptericins, defensins, ponericins, drosomycin, metchnikowin, lebocins, dipterins, and jelleines [3]), new molecules continue to be discovered and need to be characterized.

Therefore, it is fundamental to examine the in vitro biological properties of these novel compounds by evaluating (1) their antimicrobial activity and (2) their potential cytotoxic effects toward mammalian cells.

A brief description of the in vitro assays used to these purposes is reported below:

1. To perform a rapid and accurate screening of antimicrobial activity of several AMPs, the inhibition zone assay is one of the most used methods [8]. By this assay, it is possible to evaluate whether a peptide is endowed with an antimicrobial activity and if this is more selective for Gram-positive or Gram-negative bacteria. The antibacterial activity is evaluated by the inhibition zone of microbial growth that appears on an agar plate previously inoculated with the tested microorganism. The potency against Gram-positive and/or Gram-negative bacteria is related to the diameter of the inhibition halos. Peptides showing antibacterial activity can be further characterized by evaluating their minimum concentrations capable of inhibiting microbial growth in culture medium (minimal inhibitory concentration, MIC) and causing direct killing of bacterial cells (bactericidal concentration).
2. To evaluate the AMPs' effect(s) toward mammalian cells, hemolytic and cytotoxic assays can be performed.

The hemolytic assay is used to investigate the effect of AMPs on mammalian red blood cells (i.e., erythrocytes). Aliquots of a mammalian erythrocytes suspension in 0.9% NaCl are incubated with serial twofold dilutions of the peptide, and cytolytic effect is evaluated by measuring the content of hemoglobin released from damaged erythrocytes after peptide treatment.

The cytotoxic assay is carried out by a colorimetric method to assess cell viability [9]. This assay is based on the intracellular reduction of the 3-(4,5-dimethylthiazol-2yl)-2,5-diphenyltetrazolium bromide (MTT), a yellow tetrazolium salt, to a purple compound, called formazan, by mitochondrial dehydrogenases that are functional only in metabolically active cells. The MTT assay is suitable for both immortalized and primary cell lines. Here the MTT assay protocol is described by using some immortalized cell lines as examples.

2 Materials

Prepare all materials at room temperature (R.T.). Importantly, follow all waste regulations when disposing waste material.

2.1 Characterization of the Antimicrobial Activity

Reference strains of Gram-negative (e.g., *Escherichia coli* ATCC 25922) and Gram-positive bacteria (e.g., *Staphylococcus epidermidis* ATCC 12228) to be manipulated in a class I biosafety cabinet.

2.1.1 Bacterial Culture

1. Luria-Bertani (LB) broth: 1% (w:v) bacto tryptone, 0.5% (w:v) yeast extract, and 0.5% (w:v) NaCl in distilled water. The final pH has to be adjusted to 7.4 with 1 N sodium hydroxide (*see* **Note 1**). Using a graduated cylinder, dispense aliquots into glass bottles and autoclave them at 120 °C for 20 min. Store the sterilized medium at R.T.
2. LB agar Petri dishes: add 1.5% (w:v) agar to freshly prepared LB medium and autoclave at 120 °C for 20 min (*see* **Notes 2** and **3**).
3. 50-mL BD Falcon polypropylene tubes.
4. 10-μL sterile inoculating loops.
5. Spectrophotometer.
6. 1.5-mL plastic cuvettes.

2.1.2 Inhibition Zone Assay

1. 1% LB agarose: add 1% (w:v) agarose to freshly prepared LB medium and autoclave at 120 °C for 20 min. Store the sterilized medium at R.T.
2. 15-mL BD Falcon polypropylene tubes.
3. Pipette/micropipettes with disposable tips.
4. 90-mm Petri dish plates.
5. Vacuum pump.
6. Disposable glass Pasteur pipettes.
7. Adjustable work surface.
8. Scientific ruler.
9. Microwave.

2.1.3 Minimal Inhibitory Concentration (MIC)

1. Sterile 96-well polystyrene transparent flat bottom plates.
2. Sterile ultrapure water.
3. Pipette/micropipettes with disposable tips.
4. Mueller-Hinton (MH) broth: 0.2% (w:v) beef extract powder, 1.75% (w:v) acid digest of casein, and 0.15% (w:v) soluble starch in distilled water (*see* **Note 1**). Using a graduated cylinder, dispense aliquots into glass bottles and autoclave them at 121 °C for 15 min. Store the sterilized medium at R.T.
5. Plate sealer (i.e., parafilm).
6. 37 °C incubator.

2.1.4 Bactericidal Concentration (MBC) Determination

1. LB soft agar: add 0.75% (w:v) agar to freshly prepared LB medium and autoclave at 120 °C for 20 min.
2. 90-mm Petri dishes.
3. Pipette/micropipettes with disposable tips.
4. 1.5-mL microcentrifuge tubes.
5. Thermomixer.
6. Phosphate-buffered saline without calcium and magnesium chloride (CMF-PBS 1×).
7. 37 °C incubator.
8. Colonies counter.
9. Microwave.

2.2 Characterization of the Cytotoxic Effect of AMPs

2.2.1 Hemolytic Assay

1. Defibrinated mammalian blood.
2. Pipette/micropipettes with disposable tips.
3. Plastic Pasteur pipettes.
4. 0.9% (w:v) NaCl.
5. Spectrophotometer.
6. Thermomixer.
7. Centrifuge (for 50-mL tubes) and microcentrifuge.
8. 1.5-mL microcentrifuge tubes.
9. 1.5-mL plastic cuvettes.
10. 15-mL BD Falcon polypropylene tubes.
11. Distilled H_2O.

2.2.2 Cell Culture, Passaging, and Culture Media

1. 5% CO_2 incubator for cell cultures.
2. Tissue culture-treated T25 and T75 flasks (25 cm^2 and 75 cm^2, respectively).
3. Immortalized cell lines: human keratinocytes (HaCaT), human type II alveolar epithelial cells (A549) to manipulate under a class I biosafety cabinet, murine macrophages (RAW

Table 1
Differences between components of culture media for the different cell lines described

		Complete cell culture medium components					
		Cell culture medium	L-Glutamine (mM)	FBS (%)	Penicillin/ streptomycin (mg/mL)	Sodium pyruvate (mM)	NEAA (mM)
Cell line	HaCaT	DMEM	4	10	0.1	–	–
	A549	DMEM	2	10	0.1	–	–
	RAW 264.7	DMEM	2	10	0.1	1	1

264.7) to manipulate under a class II biosafety cabinet (Table 1).

4. Cell culture medium: Dulbecco's modified Eagle's medium (DMEM).
5. Heat-inactivated fetal bovine serum (FBS) (*see* **Note 4**).
6. Amino acids: L-glutamine, nonessential amino acids (NEAA).
7. Antibiotics (penicillin/streptomycin).
8. Supplements: sodium pyruvate.
9. Phosphate-buffered saline without calcium and magnesium chloride (CMF-PBS): stock solution (10×) is prepared by mixing 1.37 M NaCl, 27 mM KCl, 100 mM Na_2HPO_4, and 20 mM KH_2PO_4 in distilled water; adjust pH to 7.4 with HCl if necessary. Autoclave at 120 °C for 20 min before storage at R.T. Working buffer is prepared by diluting one part of the stock solution with nine parts of distilled water (CMF-PBS 1×). Store at R.T.
10. Cell dissociation reagent: 0.25% trypsin or trypsin/EDTA (*see* **Note 5**).
11. Cell scrapers 1.8-cm blade.
12. 50-mL BD Falcon polypropylene tubes.
13. Centrifuge for 50-mL tubes.
14. Pipette/micropipettes with disposable tips.
15. Inverted optical microscope.
16. Burker or Neubauer chamber.
17. Cover glasses.

2.2.3 MTT Assay

1. Sterile 96-well polystyrene flat bottom and tissue culture-treated transparent plates.
2. Pipette/micropipettes with sterile disposable tips.

3. Multichannel pipette to measure volumes ranging from 20 to 200 μL.
4. 1.5-mL microcentrifuge tubes.
5. Hank's balanced salt solution (HBSS): 136 mM NaCl; 4.2 mM Na_2HPO_4; 4.4 mM KH_2PO_4; 5.4 mM KCl; 4.1 mM $NaHCO_3$, pH 7.2, supplemented with 20 mM D-glucose.
6. MTT stock solution: use sterile HBSS to dissolve MTT to a concentration of 5 mg/mL, and store single-use aliquots at −20 °C in dark or foil-covered bottles, since the compound is light sensitive. Prepare fresh working solutions at the final desired concentration (0.5 mg/mL) by dilution with HBSS (*see* **Note 6**).
7. Stop solution: prepare it by adding 0.04 N HCl to isopropanol.
8. A microplate reader for absorbance measurements at 570 nm.
9. Aluminum foil.
10. Plate sealer (i.e., parafilm).

3 Methods

3.1 Characterization of the Antimicrobial Activity

3.1.1 Inhibition Zone Assay

1. Under the biological safety cabinet, open a tube of frozen glycerol stock containing bacterial cells. Scrape off a portion from the top of the frozen glycerol with a 10-μL sterile loop and streak it onto an LB agar plate.
2. In the case of the Gram-negative bacterium *E. coli* or the Gram-positive bacterium *S. epidermidis*, the plates are incubated at 37 °C for 24 h under aerobic conditions.
3. Pick a single bacterial colony from the plate, and inoculate it into ~10 mL LB broth, previously aliquoted into a 50-mL polypropylene Falcon tube. Incubate the inoculated broth at 37 °C (use an incubator with shaking at 150 rpm), until an optical density $(OD)_{590}$ = 0.8 is reached.
4. When the microbial culture has reached the correct OD, dilute as follows: for *E. coli* take 100 μL and dilute in 4.9 mL of LB, while for *S. epidermidis* take 300 μL and dilute in 4.7 mL of LB.
5. Melt the 1% LB agarose in a microwave and aliquot 6 mL in 15-mL polypropylene Falcon tube (*see* **Note 7**).
6. Add 50 μL of the freshly diluted bacterial culture into the 6 mL of lukewarm LB agarose. Shake the tube to mix the bacteria, and pour the content into a 90-mm Petri dish, previously laid on an adjustable work surface in a perfectly horizontal position (*see* **Note 8**).

7. Allow the agarose medium to solidify and dry for about 30 min.
8. Prepare the stock solution of the test compounds (*see* **Note 9**).
9. Once the medium has solidified, place a sheet of paper showing the position of holes that have to be made in the solidified medium, below the plate (*see* Fig. 1a). Using a P10 plastic tip, make holes at the circumferences drawn on the diagram.
10. With the help of disposable glass Pasteur pipettes and a vacuum pump, aspirate the agarose medium from the newly created circles (*see* **Note 10**).
11. Inoculate 3 μL of each test compound in one or more holes, depending on the number of compounds (*see* **Note 11**).
12. Allow the drop of sample to diffuse in the LB agarose for 30 min, then seal the plate with parafilm without shaking, and incubate it overnight at 30 °C, with the bottom facing down.
13. After incubation of about 16–18 h, observe the halos of inhibition of microbial growth as shown in Fig. 1b. Measure the diameter of the halos with a scientific ruler. The larger the diameter of the halo, the more efficient is the tested compound. Based on the presence/absence of inhibition halos in the plates seeded with Gram-positive or Gram-negative bacteria, a target cell selectivity can be deduced.

3.1.2 Minimal Inhibitory Concentration (MIC) Determination

This assay serves to determine the minimal concentration of antimicrobial peptide capable of inhibiting the microbial growth in liquid broth. It is based on the measurement of turbidity that the culture medium acquires when bacteria grow without limitations, and it is commonly named as "microdilution broth method."

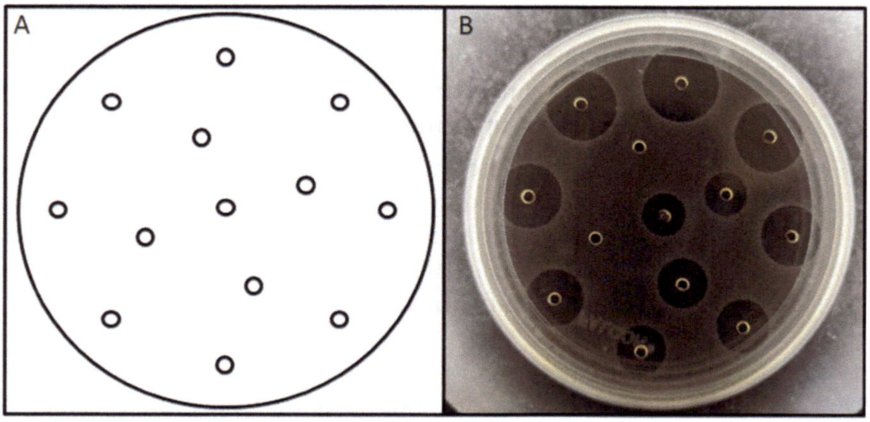

Fig. 1 (**a**) Schematic representation of a Petri dish and multiple holes for peptide inoculation. (**b**) Representative inhibition zone assay in a Petri dish. The test compounds are inoculated at the center of the holes produced in the agarose plate. Transparent halos around the holes indicate the capability of the inoculated peptides to inhibit microbial growth

1. Under the biological safety cabinet class, open a tube of frozen glycerol stock containing bacterial cells. Scrape off a portion from the top of the frozen glycerol with a 10-μL sterile loop and streak it onto an LB agar plate.
2. In the case of the Gram-negative *E. coli* or the Gram-positive *S. epidermidis* bacteria, the plates are incubated at 37 °C for 24 h under aerobic conditions.
3. Pick a single bacterial colony from the plate, and inoculate it into ~10 mL LB broth, previously aliquoted into a 50-mL polypropylene Falcon tube. Incubate the inoculated broth at 37 °C (use an incubator with shaking at 150 rpm), until an OD_{590} = 0.8 is reached.
4. When the microbial culture has reached the correct OD, dilute as follows: for *E. coli* take 25 μL and dilute in 4.975 mL of MH, while for *S. epidermidis* take 50 μL and dilute in 4.950 mL of MH.
5. Prepare a stock (5 mL) of 90% MH diluted in vehicle of peptides.
6. Open a sterile 96-well transparent flat bottom and prepare it as follows, taking into account that in each column or row a compound can be tested in serial dilutions: put 90 μL of MH in the first well of each column.
7. Put 50 μL of 90% MH from the second well to the last one.
8. Inoculate 10 μL of the tested compound in the first well of each column (*see* **Note 12**).
9. Take 50 μL from the first well and dilute them in the second; then, take 50 μL from the second well and dilute them in the third one and so on. After the last well, discard the remaining 50 μL. At the end, each column should have 50 μL of medium containing serial twofold dilutions of peptides.
10. Place 50 μL of bacterial culture (prepared in **step 4**) in each well. Controls are represented by 50 μL of 90% MH without peptides supplemented with 50 μL of bacteria.
11. Seal the plate with parafilm and incubate it in a 37 °C incubator with shaking at 150 rpm for 18 h. The minimal inhibitory concentration (MIC) is defined as the concentration of peptide at which 100% inhibition of microbial growth is visually observed after incubation time.

3.1.3 Bactericidal Concentration Determination

This assay allows to evaluate the killing capacity of the compound by determining the concentration of peptide capable of reducing the number of viable cells. It is a quantitative assay based on colony-forming units (CFU) counting (*see* **Note 13**).

1. Under the biological safety cabinet, open a tube of frozen glycerol stock containing bacterial cells. Scrape off a portion from the top of the frozen glycerol with a 10-μL sterile loop and streak it onto an LB agar plate.
2. In the case of the Gram-negative *E. coli* or the Gram-positive *S. epidermidis* bacteria, the plates are incubated at 37 °C for 24 h under aerobic conditions.
3. Pick a single bacterial colony from the plate, and inoculate it into ~10 mL LB broth, previously aliquoted into a 50-mL polypropylene Falcon tube. Incubate the inoculated broth at 37 °C (use an incubator with shaking at 150 rpm), until an $OD_{590} = 0.8$ is reached (*see* **Note 14**).
4. Centrifuge the tube for 5 min at 1400 × *g*, discard the supernatant, and wash the pellet with ~5 mL of CMF-PBS 1×.
5. Centrifuge again, for 5 min at 1400 × *g*, discard the supernatant, and add ~10 mL of CMF-PBS 1× to reach an OD_{590} of 0.8.
6. For *E. coli*, prepare a diluted bacterial culture (5 mL, 1×10^6 CFU/mL) by adding 13 μL of the bacterial cell suspension (at $OD_{590} = 0.8$) in 4987 μL of PBS. For *S. epidermidis*, prepare a diluted culture (5 mL) by adding 26 μL of the bacterial cell suspension (at $OD_{590} = 0.8$) in 4974 μL of PBS.
7. Prepare several 1.5-mL microcentrifuge tubes depending on the number of peptide concentrations that have to be tested. Remember to add one more tube for the vehicle control.
8. In each microcentrifuge tube, add 95 μL of the diluted microbial culture.
9. Prepare the test peptides at different stock concentrations in microcentrifuge tubes. A concentration 20× higher than the desired final dosage is recommended.
10. Melt the LB soft agar with a microwave.
11. Add 5 μL of each peptide concentration in the corresponding microcentrifuge tube.
12. Samples are incubated in the thermomixer, with shaking at 600 rpm, at 37 °C for different time points (e.g., 30, 90, and 120 min; *see* **Note 15**).
13. At the corresponding time intervals, 5 μL aliquots of treated samples are aspirated, diluted in 6 mL of LB soft agar, and plated on LB agar plates. For controls, a 1:10 dilution is made before plating.
14. Incubate the plates overnight at 37 °C (*see* **Note 16**).
15. Count the number of CFU and express the results as percentage with respect to the untreated ones (*see* **Note 17**).

3.2 Characterization of the Cytotoxic Effect of AMPs

3.2.1 Hemolytic Assay

1. Centrifuge 10 mL of blood for 10 min at 1400 × *g* to separate the plasma (supernatant) from the corpuscular part (pellet of red blood cells) (*see* **Note 18**).
2. Aspirate the supernatant with a plastic Pasteur pipette (*see* **Note 19**).
3. Rinse freshly drawn erythrocytes twice with ~10 mL of 0.9% NaCl, and centrifuge samples for 10 min at 1400 × *g*.
4. Resuspend erythrocytes with 5–10 mL of 0.9% NaCl, and split the cell suspension in two different aliquots with the same volume. To perform the assay, use only one of these aliquots (erythrocytes stock solution).
5. Aspirate 600 μL of erythrocytes stock solution; add them to 5.4 mL of distilled H_2O (lysate solution) and measure the absorbance at 500 nm (*see* **Note 20**).
6. The OD of the lysate solution should be 0.45–0.5 which corresponds to a cell density of 1×10^7 erythrocytes/mL (*see* **Note 21**).
7. Store the lysate solution with the correct OD and use it as a control (*see* **Note 22**).
8. Prepare the working solution by adding 600 μL of erythrocytes stock solution to 5.4 mL of 0.9% NaCl.
9. Prepare the test peptides at a concentration 20× higher than the desired final dosage.
10. In each microcentrifuge tube, add 95 μL of the working solution and 5 μL of AMP.

    For blank controls add 5 μL of the peptide vehicle.

    For the positive control (total lysis), add 95 μL of lysate solution +5 μL of the peptide vehicle (*see* **Note 23**).
11. Incubate samples at 37 °C in the thermomixer with shaking at 500 rpm for 30 min (*see* **Note 24**).
12. Centrifuge samples at 900 × *g* for 5 min. After centrifugation, pellets and supernatants of the samples should appear as reported in Fig. 2.
13. Remove the supernatant (90 μL) being careful not to aspirate the pellet, and transfer it into a 96-well plate (*see* **Note 25**).
14. Measure the absorbance at 540 nm and 415 nm.
15. Calculate the mean value of the blank and positive controls, and then calculate the percentage of hemolysis according to the formula:

$$\frac{\left(\text{absorbance}_{\text{sample}} - \text{absorbance}_{\text{blank control}}\right)}{\left(\text{absorbance}_{\text{total lysis control}} - \text{absorbance}_{\text{blank control}}\right)} \times 100$$

    The positive control corresponds to 100% hemolysis.
16. Plot graph.

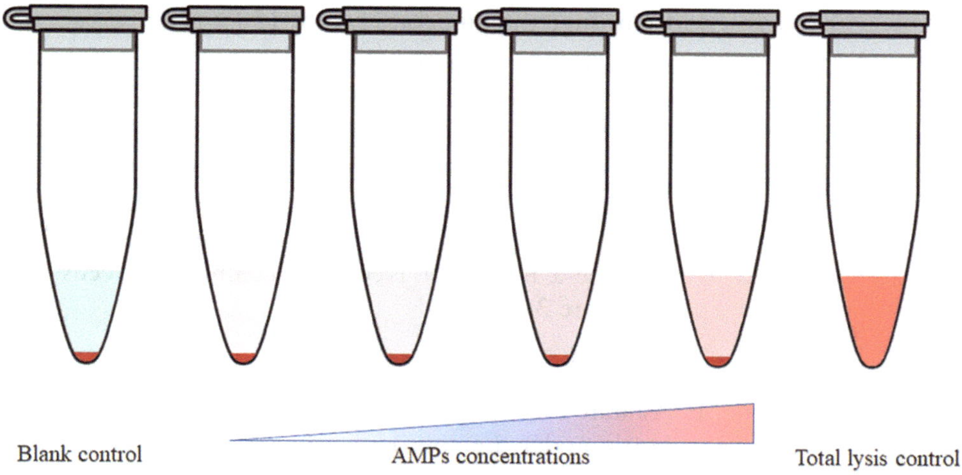

Fig. 2 Representative 1.5-mL microcentrifuge tubes containing erythrocytes and different concentrations of AMPs, after centrifugation. The hemolytic effect is calculated by evaluating the amount of released hemoglobin by absorbance measurement of samples after centrifugation. Blank control (vehicle-treated sample) and total lysis control (erythrocytes in distilled water) are shown for comparison

3.2.2 Cell Culture, Passaging, and Culture Media

The cell lines are generally cultured in T25 or T75 flasks using appropriate complete cell culture media (Table 1) that allow cell growth (generally, 5 mL in a T25 flask, 10 mL in a T75 flask). The cells are incubated at 37 °C and 5% CO_2. When a 90–100% cell confluence is reached, detach cells from the flask as described below (*see* **Note 26**). All steps must be carried out under the biosafety cabinet to maintain sterile conditions.

For HaCaT and A549 cell lines:

1. Aspirate the medium from the flask and discard it into a waste bottle under the biosafety cabinet. Perform two fast washes with CMF-PBS 1× (4 mL in a T25 flask, 6 mL in a T75 flask). For each wash, gently shake the flask manually and discard CMF-PBS 1×. Add the same volume of CMF-PBS 1× for 3 min under the biological safety cabinet, and then discard it (*see* **Note 27**).
2. Add an appropriate volume of 0.25% trypsin or trypsin/EDTA to the flask (i.e., 1 mL in a T25 flask, 2 mL in a T75 flask). Gently shake the flask, allowing the solution to completely coat the cells, and incubate the flask at 37 °C and 5% CO_2. Check that the cells are visibly detached under an inverted microscope. Generally, it takes at least 8 min and 5 min for HaCaT and A549 cells, respectively. However, it is recommended to observe cells every few min intervals (*see* **Note 28**).
3. Add DMEM supplemented with glutamine (DMEMg, at the concentration indicated in Table 1) and 10% FBS (6 mL in a T25 flask, 10 mL in a T75 flask) to inactivate trypsin. Collect and transfer the cells into a conical 50-mL tube.

4. Centrifuge the tube for 5 min at 150 × *g*.
5. Aspirate the supernatant, and then resuspend the cells in DMEMg supplemented with 2% FBS (i.e., 3 mL for a T25 flask, 5 mL for a T75 flask). Only for HaCaT cells use a syringe with needle to break up any clumps that may be present (*see* **Note 29**).
6. Aspirate 10 μL of cell suspension with a micropipette tip, and use a Burker or Neubauer chamber for cells counting (*see* **Note 30**).

For RAW 264.7 cell line:

1. Aspirate the medium from the flask and discard it into a waste bottle under the appropriate biosafety cabinet. Perform two fast washes with CMF-PBS 1× (4 mL in a T25 flask, 6 mL in a T75 flask). For each wash, gently shake the flask manually and discard CMF-PBS 1×. Add the same volume of CMF-PBS 1× for 3 min under the biological safety cabinet, and then discard it (*see* **Note 27**).
2. Add an appropriate volume of cell culture medium DMEM containing glutamine (DMEMg supplemented with sodium pyruvate and NEAA, at the concentration indicated in Table 1) and with 2% FBS (i.e., 4 mL in a T25 flask, 8 mL in a T75 flask). Mechanically and gently detach the cells using the cell scraper, and check that the cells are visibly detached under an inverted microscope (*see* **Note 28**).
3. Collect and transfer the cells into a conical 50-mL tube.
4. Aspirate 10 μL of cell suspension with a micropipette tip for counting (*see* **Note 30**).

3.2.3 MTT Assay

1. Seed 40,000 cells (resuspended in DMEMg or DMEMg supplemented with sodium pyruvate and NEAA, plus 2% FBS) in a volume of 100 μL for each well of a 96-well plate. Incubate the plate overnight (O.N.) at 37 °C and 5% CO_2 to allow the cells to attach and grow to confluence.
2. Check that cells have reached the confluence.
3. Resuspend the test compounds (AMPs) in the proper cell culture medium without FBS, to avoid any chemical interactions with the test compounds (*see* **Note 31**).
4. Aspirate the medium from each well under the biosafety cabinet (*see* **Note 32**).
5. Wash the cells with 100 μL/well of fresh serum-free medium.
6. Add the test compounds (prepared at **step 3**) to the wells (100 μL/well). For untreated control samples, add 100 μL of serum-free medium containing peptide vehicle at the same concentrations of the treated samples. Incubate the plate at 37 °C and 5% CO_2 for the desired time, generally 2 h or 24 h.

7. Aspirate the medium from each well and wash the cells with 100 μL/well of HBSS (*see* **Note 32**).
8. In the darkness, add 100 μL MTT solution at final concentration of 0.5 mg/mL into each well (*see* **Note 33**). For blank samples, add MTT solution in at least three wells without cells. Incubate the plate at 37 °C and 5% CO_2 for 4 h to allow the formation of purple formazan crystals.
9. Stop the reaction by adding 100 μL of acidified isopropanol to solubilize the formazan crystals. Seal the plate with parafilm and cover with aluminum foil (*see* **Note 34**).
10. Read the absorbance of the wells using a multiplate reader at 570 nm.
11. Calculate the mean value of untreated control samples and blank samples; then calculate the percentage of metabolically active cells (i.e., viable cells), according to the formula:

$$\frac{\left(\text{absorbance}_{\text{sample}} - \text{absorbance}_{\text{blank}}\right)}{\left(\text{absorbance}_{\text{control}} - \text{absorbance}_{\text{blank}}\right)} \times 100$$

where the blank is given by samples without cells and not treated with the peptide.
12. Plot graph.

4 Notes

1. Different culture media and growth media are used for different types of bacterial strains and assays, according to standard microbiology procedures.
2. Be careful to cool the medium to 45–50 °C prior to pouring the plates. This is to minimize the condensation of water vapor that may be formed.
3. Add 15–20 mL of medium into 90-mm dish plates and store them at +4 °C.
4. If FBS is not inactivated, proceed with heat inactivation for 30 min at 56 °C to destroy the complement molecules and the immunoglobulins that may be present to avoid a cascade of complement reaction and final cell lysis.
5. It is preferable that the trypsin is in a balanced saline solution free of calcium and magnesium ions (CMF-PBS). In fact, these ions usually mask the bonds on which the trypsin acts and increase the cell adhesion. Other cell dissociation reagents can be used, for example, TrypLE Express. This latter does not require the use of media containing 10% FBS to stop the trypsin reaction. However, it is necessary to carefully evaluate

the incubation times of each cell dissociation reagent to avoid that its effect is too strong and harmful to the cells.

6. Since MTT is toxic, use personal protective equipments, such as eyeshields, gloves, and respirator filter. MTT is also light-sensitive; therefore, weigh the compound in the darkness. Then, solubilize it in HBSS and shake it with a vortex mixer. Sterilize the obtained MTT solution in a biological safety cabin by a sterile filtration device fitted with a 0.22 μm nitrocellulose filter. Fresh working solution can be stored at +4 °C for a few days.
7. Make sure that temperature is not too high (it should be around 45–50 °C).
8. Make sure that the 1% LB agarose evenly covers the surface of the plate with an equal thickness at each point of the plate.
9. Make sure the concentration is high enough to obtain a visible result, e.g., 1–2 mM.
10. Take care not to damage the edge of the holes and make sure that they are as precise as possible.
11. Make sure that the drop does not escape from the hole. The drop must remain within the edges of the circle.
12. Note that for a good range of concentrations, the stock solution of the tested compound should have a concentration of at least 1 mM.
13. The assay should be designed on the basis of the characteristics of the compound under examination. The choice of the buffer, the incubation time, and the dilution of bacterial culture for CFU counting can vary among different experiments depending on the desired experimental conditions.
14. At this OD, the concentration of *E. coli* and *S. epidermidis* cells should be around 4×10^8 CFU/mL and 2×10^8 CFU/mL, respectively.
15. The time points may vary based on the mechanism of action of the peptide. For peptides with fast killing kinetics, short time intervals should be chosen. For slower kinetics, longer time intervals are recommended.
16. Leave the plate drying at R.T. for 10–20 min. Then invert the plate and incubate as desired.
17. A proper number of countable bacterial cells is between 50 and 500 CFU per plate.
18. After centrifugation, a visible white disk (leukocytes) appears between plasma and corpuscular part.
19. Remove also the leukocytes disk with the supernatant being careful not to aspirate the corpuscular part.

20. To obtain the right concentration of erythrocytes, measure the absorbance at 500 nm of hemoglobin released from lysed erythrocytes in distilled water at 50 °C.
21. If necessary, adjust the concentration of the erythrocytes stock solution by adding 0.9% NaCl. Repeat the measurement of OD of the lysate solution until the correct OD is reached.
22. If 95 μL of lysate are centrifuged at 900 × *g* for 5 min, no pellets should be obtained.
23. Both controls should be run in triplicates.
24. It is possible to evaluate the hemolytic effect at different time points.
25. After centrifugation the blank control should be clear and should have a pellet, while the positive control should be red without any pellet.
26. Each cell line has its own characteristics that require the use of specific components of the cell culture medium and specific steps for cell passaging.
27. Be careful not to touch the cell monolayer with the pipette during washes, to avoid breaking the cell monolayer.
28. When the cells are detached from each other and from the surface, they appear rounded. If the cells are not well detached, soft manual agitation may be necessary or, in the case of RAW 264.7 cell line, continue to gently scrape the cells from the flask.
29. The passage of the cells through the needle must be done quite slowly to avoid cells rupture.
30. Gently turn the cell suspension to ensure that the cells are evenly distributed and aspirate 10 μL. Slowly release this volume under the glass cover edge, previously put above the Burker or Neubauer chamber. The liquid is absorbed by capillarity; therefore, control that it enters the chamber uniformly. In case of bubble formation, rinse the chamber and repeat the loading process. If the cell suspension is too concentrated, dilute it. Count the cells visible in two squares (each one contains 16 small squares, bounded by triple lines in Burker chamber or multiple lines in Neubauer chamber). Repeat the operation for the other side of the chamber. Calculate the mean value (number of cells/number of squares), and multiply it by 10,000 to obtain the concentration of cell suspension (cells/mL). Any applied dilution must be considered in the calculation.
31. Firstly, prepare fresh AMPs dilutions starting from the stock solution stored at −20 °C, at a concentration 12.5× higher than the desired final dosage. Then, add 28 μL of these dilutions to 322 μL of DMEMg. This volume is sufficient for the triplicates.

32. If cells are seeded in many wells, use a multichannel pipette to avoid too long dry time during washes. Be careful not to touch the cell monolayer with the tip during all the steps, to avoid breaking the cell monolayer.
33. Dilute the stock MTT (5 mg/mL) in HBSS to the final desired concentration.
34. Since full solubilization may take a few hours, incubate the plate at 37 °C and soft agitation for 1 h to speed up the process.

References

1. Sierra JM, Fuste E, Rabanal F, Vinuesa T, Vinas M (2017) An overview of antimicrobial peptides and the latest advances in their development. Expert Opin Biol Ther 17(6):663–676. https://doi.org/10.1080/14712598.2017.1315402
2. Ageitos JM, Sanchez-Perez A, Calo-Mata P, Villa TG (2017) Antimicrobial peptides (AMPs): ancient compounds that represent novel weapons in the fight against bacteria. Biochem Pharmacol 133:117–138. https://doi.org/10.1016/j.bcp.2016.09.018
3. Mylonakis E, Podsiadlowski L, Muhammed M, Vilcinskas A (2016) Diversity, evolution and medical applications of insect antimicrobial peptides. Philos Trans R Soc Lond Ser B Biol Sci 371(1695):20150290. https://doi.org/10.1098/rstb.2015.0290
4. Wu Q, Patocka J, Kuca K (2018) Insect antimicrobial peptides, a mini review. Toxins (Basel) 10(11):E461. https://doi.org/10.3390/toxins10110461
5. Yi HY, Chowdhury M, Huang YD, Yu XQ (2014) Insect antimicrobial peptides and their applications. Appl Microbiol Biotechnol 98(13):5807–5822. https://doi.org/10.1007/s00253-014-5792-6
6. Casciaro B, Loffredo MR, Luca V, Verrusio W, Cacciafesta M, Mangoni ML (2018) Esculentin-1a derived antipseudomonal peptides: limited induction of resistance and synergy with aztreonam. Protein Pept Lett 25(12):1155–1162. https://doi.org/10.2174/0929866525666181101104649
7. Ghosh C, Sarkar P, Issa R, Haldar J (2019) Alternatives to conventional antibiotics in the era of antimicrobial resistance. Trends Microbiol 27(4):323–338. https://doi.org/10.1016/j.tim.2018.12.010
8. Balouiri M, Sadiki M, Ibnsouda SK (2016) Methods for in vitro evaluating antimicrobial activity: a review. J Pharm Anal 6(2):71–79. https://doi.org/10.1016/j.jpha.2015.11.005
9. Riss TL, Moravec RA, Niles AL, Duellman S, Benink HA, Worzella TJ, Minor L (2004) Cell viability assays. In: Sittampalam GS, Coussens NP, Brimacombe K et al (eds) Assay guidance manual. Eli Lilly and the National Center for Advancing Translational Sciences, Bethesda (MD)

Chapter 10

Characterization of Antimicrobial Peptide–Membrane Interaction Using All-Atom Molecular Dynamic Simulation

Shruti Mukherjee, Rajiv K. Kar, and Anirban Bhunia

Abstract

The investigation of peptides interaction with cell membranes is essential for understanding the basic functions such as membrane transport, fusion, and signaling processes, which may elucidate the potential applications of peptides in biomedicine. Antimicrobial peptides (AMPs) are now widely explored as an alternative to antibiotics owing to their superior ability to disrupt cell membrane both alone and with cargo molecules. Understanding the interaction mechanism of AMP is significant for many therapeutic purposes, including targeted microbial cell death. Cell membranes are mostly characterized by a membrane bilayer, which presents a complex and heterogeneous association of molecules flexible for an external agent. Hence, studies of protein–membrane interactions constitute a challenge in the structural biology field. Molecular dynamics (MD) is one of the useful methods to investigate membrane-associated processes. An extensive set of model bilayers and micelles differing in lipid composition are used to study different classes of membrane-active proteins and peptides like toxins, antimicrobial, Trojan, and fusion peptides. Regardless of the limitations of the MD timescale, membrane simulation results are capable of giving a balanced picture for the mechanism of action.

Key words MD simulation, Antimicrobial peptides, Cell membrane, CecB

1 Introduction

Antimicrobial peptides (AMPs) have been discovered in all classes of life, ranging from bacteria to mammals—as a protective mechanism against infection. Many of these AMPs have been isolated and extensively studied to elucidate the molecular mechanism of their mode of action. Though the mechanisms are still being argued in the scientific community, essential clues have been found through structure–function relationship studies. MD simulation is one of the useful tools in the context of a biophysical and biochemical investigation of protein, nucleic acid, small molecule, and membranes [1, 2]. In this chapter, we will illustrate the usefulness of MD simulation technique for the characterization of AMP–membrane interaction, by taking a case example of the *Bombyx mori* Cecropin B (CecB E53 and Q53 isoforms) in a Gram-negative

Federica Sandrelli and Gianluca Tettamanti (eds.), *Immunity in Insects*, Springer Protocols Handbooks,
https://doi.org/10.1007/978-1-0716-0259-1_10,

bacterial membrane model [3, 4], composed of 1-palmitoyl-2-oleoyl-*sn*-glycero-3-phosphoethanolamine (POPE) and 1-palmito yl-2-oleoyl-*sn*-glycero-3-phosphoglycerol (POPG) in 3:1 ratio.

1.1 Few Known AMPs

Cecropin is the first insect antimicrobial peptide purified from *Hyalophora cecropia* in 1984 [5], and till date, more than 150 AMPs have been identified with known activity. The AMPs sourced from insects are mainly comprised of short peptide fragments, which are charged in nature and active against a wide range of bacteria and fungi [6]. Despite diverse secondary structure, the AMPs undergo structural transition, forming folded secondary structure (α-helices, β-sheets, or coil) upon interaction with the microbial membrane system [7]. Previous studies have suggested that the dynamic conformations of these peptides are crucial in governing the mechanism of action and specific cytotoxicity [8, 9]. It is important to mention that the characteristics of AMPs, such as charge, amphipathicity, and secondary structure, are key factors, which mediates their antimicrobial action [10].

1.2 The Usefulness of MD Simulation

Elucidating the three-dimensional structure of proteins is one of the critical objectives in the field of structure-based drug design. These structural coordinates are the starting point for MD simulation studies. However, structures present in the Protein Data Bank (PDB) [11] provide only limited details of conformation and flexibility, which can be accessed through simulations. For example, (1) AMPs are flexible entities, and dynamics can play a crucial role in their functionality, (2) AMPs undergo significant conformational changes while performing their function, and also (3) the influence of the conformation on macromolecular function cannot be revealed through rigid structures. Nevertheless, the recent developments in molecular dynamics (MD) simulations [12] are highly relevant for elucidating the structure–function relationships in proteins, by computing reaction coordinates (conformational hyperspace) concerning timescale.

Using MD simulations, detailed atomistic information can be obtained, which might be challenging to achieve by experimental techniques. The advancement in computational power allows to access the experimentally relevant timescales and investigate the mechanism such as peptide binding to biomacromolecules, folding, and partitioning into lipid bilayers. Further, it helps in understanding how the polypeptide fragments form spontaneous channels, facilitate ion exchange or proton pump [13], and mediate delivery of cargo molecules across membranes [14].

1.3 Choosing Appropriate Force Field for Simulations

One of the relevant variables used in the MD simulations, to study the protein–lipid interaction, is the selection of appropriate molecular mechanics force field. It is a well-known fact that different force fields (AMBER, CHARMM, and OPLS) have varying parameters for proteins and lipids, which are based on their development schemes (level of theory used for parameterization) [15]. Also advances in computing algorithm, including parallel architectures, and efficient codes such as NAMD [16], have made significant contributions to mimic the biological interactions through the physics-based force fields and an explicit representation of water molecules. In the field of molecular simulation, the development of the CHARMM force field is a significant contribution, which is reflective with benchmark studies [17, 18]. Apart from proteins, CHARMM also supports nucleic acids and lipids, lipopolysaccharides (LPS), to study biologically relevant system.

1.4 Accessing Relevant Timescale for Simulation Runs

Classical simulation is efficient in performing the time-dependent sampling of biomolecular conformations; however, the technique is limited due to the computational cost. In contrary, the relevant biological processes tend to happen on milliseconds timescale or beyond [19]. Advanced sampling techniques such as hyperdynamics [20], metadynamics [21], conformational flooding [22], and the adaptive biasing force method [23] have been developed to investigate these long timescales. Accelerated molecular dynamics (aMD) is another improved sampling technique, which aids in a rigorous sampling of events by decreasing energy barriers between high-energy states. The relevant timescale for accessing MD events typically ranges from tens to thousand nanoseconds. However, it is more important to test the convergence of trajectory and evaluate the quality of data. Also, assessment of conformational dynamicity through multiple trajectories is an appropriate method [24] often used in our group. Approaches like principal component analysis [24], transient-state analysis through the correlation of variables (RMSD or Rg), and Markov-state modeling [25] are highly relevant in this context.

1.5 Outline of this Chapter

In this book chapter, we will be explaining the methodological details used for studying peptide–membrane interaction using MD simulation. First, we will discuss the construction of the peptide–lipid system, and important notes for MD simulation will be described as a part of the protocol. Next, the method used for the MD simulation of the interaction between CecB (E53 and Q53) and 3:1 POPE/POPG lipid bilayer will be explained. We chose the particular CecB antimicrobial peptide from *Bombyx mori,* as the Cecropin family of AMP constitutes a significant part of the insect immunity. CecB is a naturally occurring linear cationic peptide consisting of 35 amino acids with the highest antibacterial activity in the silkworm Cecropin family. Further, the description will focus on the tools used for the analysis of MD events. Finally, the chapter is concluded with essential notes from the case study, practical utility, and recommendations, which might be helpful for the user community.

2 Materials

1. Modeled peptide structure using SWISS-MODEL [26] web server (https://swissmodel.expasy.org/).
2. CHARMM-GUI [27] web server to model the Gram-negative bacterial membrane as well as membrane–peptide system (http://www.charmm-gui.org/).
3. NAMD [16], a parallel molecular dynamics code developed by Theoretical and Biophysical Computational Group to simulate in the UNIX operating system. The code is downloadable from http://www.ks.uiuc.edu/Research/namd/.
4. VMD [28] and VMD membrane plugin [29] are required as the visualization and analysis tool. The software is downloadable from http://www.ks.uiuc.edu/Research/vmd/.

3 Methods

3.1 Overview of Protocol: MD Simulation with CHARMM- GUI

Reading the structural coordinates (PDB format) into a suitable format for CHARMM is supported with CHARMMing script. The system constructs for MD simulation can be prepared using CHARMM-GUI [27]. The equilibration ensemble can be kept at canonically defined (NVT), and the dynamics ensemble can be set to NPT as per criteria. It is standard to simulate under NVT conditions first to equilibrate the pressure, followed by the simulation under NPT, where pressure is fixed, yet volume is permitted to change. The measurable observables in this condition are relevant to experiments such as isothermal titration calorimetry (ITC) and include model biochemical reactions, conformational sampling, and macromolecular interaction. In this protocol, a zipped file will be available to download from the server containing all the input files to simulate with NAMD [16]. Additionally, it also provides input files which can be interfaced with other popular MD engines such as AMBER, CHARMM, Gromacs, Desmond, and OpenMM. The overall workflow is shown in Scheme 1 [30].

3.2 Structural Coordinates of the Peptide as Stating Geometry

CecB is one of the potent cationic AMPs with remarkable antimicrobial activity against *Pseudomonas aeruginosa.* In the membrane environment, CecB adopts a helical structure and perturbs the membrane architecture to mediate the action [31]. The substitution of E by Q residue at 53 positions of CecB is known to enhance the antimicrobial property of the peptide against *Pseudomonas aeruginosa* [32].

Since the coordinates of this peptide are not available in the database, we prepared a homology model, to construct the peptide–membrane system. The three-dimensional homology model of CecB (E53 and Q53 variants) was developed using the SWISS-MODEL

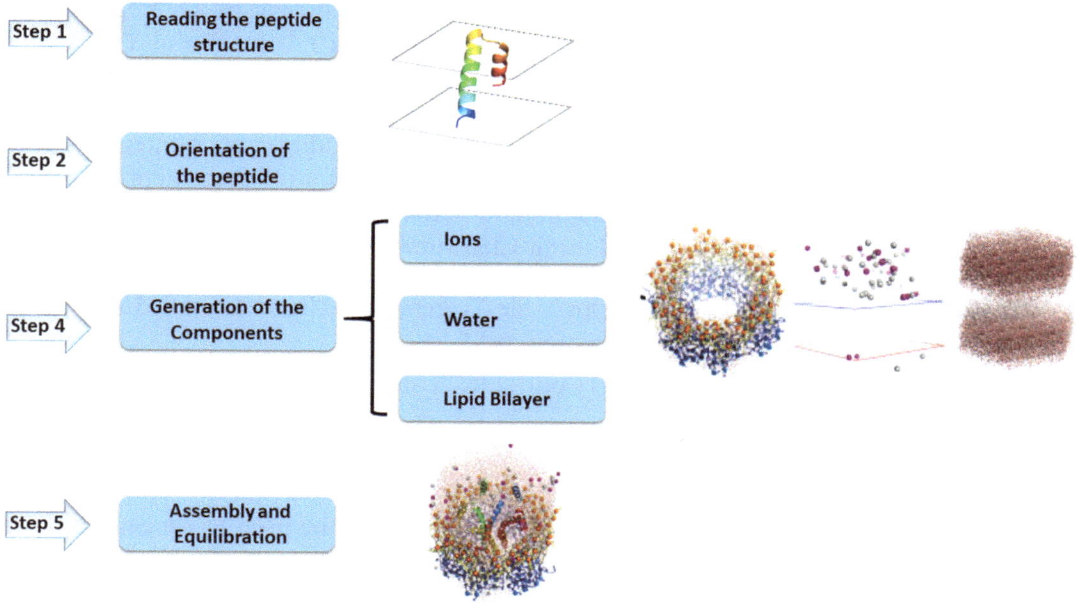

Scheme 1 Proposed workflow to model peptide–membrane interaction using molecular dynamics (MD) simulation

[26] web server. The structure of peptide papiliocin solved using NMR spectroscopy was used as a template (PDB acquisition code, 2la2.pdb) for the modeling, which has a sequence identity of 71.43%. The search for the template was performed using the BLAST algorithm. The final model was evaluated with Maestro, Schrodinger to check the close contacts, stereochemical properties, chirality, as well as bond lengths and φ–ψ angles. Other programs such as I-TASSER [33], Rosetta [34] and Medeller [35] can also be used for preparing accurate homology models.

3.3 Reading the PDB File

In this section, we will detail the processing of the PDB file, as required for using CHARMM-GUI server. (1) HIS residue name needs to be changed to HSE. (2) Subunit termini for residue numbering should be analyzed. (3) The ATOM records indicate the IUPAC (i.e., PDB) names and the CHARMM atom types for all the atoms in the residue, along with the partial atomic charges. HETATM gives the coordinates for nonstandard groups. (4) The residue numbers in the PDB record relate to RESID in CHARMM (not to be mistaken for the RESID). The RESID is treated as a name that is utilized to match with the relating molecule in the PSF.

3.4 Preparing the Membrane–Protein System

A lipid bilayer (3:1 POPE/POPG) framework was developed for this case to mimic the membrane of the Gram-negative bacterial membrane [36]. The membrane-embedded system was prepared using Membrane Builder module of CHARMM-GUI [27]. Initial positioning of the peptide inside the membrane system was determined by translating the coordinate axis by 16 Å. The membrane dimension was fixed to 20 and 40 Å, for *X* and *Y* axis, with center at $Z = 0$. Solvation box in a rectangular arrangement was prepared with edge distance of 20 Å from both leaflets. System neutralization was achieved with the addition of counter ions (Na^+ or Cl^-), using Monte Carlo technique for ion placement. The temperature of the membrane system was set to 300 K. All simulations were performed using Charmm36 force field [37] with NAMD.

3.5 Running the Simulation Using NAMD

3.5.1 Molecular System Configuration

1. Structure (.psf) file is generated from CHARMM-GUI as "structure protein.psf."
2. Coordinates (.pdb) file: "coordinates protein.PDB" (*see* **Notes 1–4**).

3.5.2 Periodic Cell Configuration

1. The following parameters should be used to set one to three basis vectors and a center point: "cellBasisVector1 60.23 0 0" "cellBasisVector2 0 46.31 0" "cellBasisVector3 0 0 30.42" "cell origin 0 0 0"
2. The restart file with "extendedSystem restart.xsc" also contains the periodic boundary informations (*see* **Note 5**).

3.5.3 Energy Function Configuration

1. Parameter file(s) "paraTypeCharmm on" is used for CHARMM force field.
2. Also, the exclusion policy is defined as "exclude scaled1-4" "1-4scaling 0.4" (*see* **Note 6**).

3.5.4 Nonbonded Cutoff Configuration

1. To define the nonbonded cutoff limit, "switching on" "switchdist 7.5" "cut off 8" "pairlistdist 9.5" "stepspercycle 10" is incorporated in the input script (*see* **Note 7**).

3.5.5 Minimization Configuration

1. To perform the conjugate gradient minimization, "minimization on" is used. Conjugate gradient minimization is an efficient, self-tuning, and very robust process for energy minimization.
2. In the case of initial instability, "minTinyStep" value should be reduced to 1.0×10^6.
3. In the case of later instability, "minBabyStep" value should be reduced to 10×10^2 (*see* **Note 8**).

3.5.6 Integrator Configuration

1. The basic configuration uses verlet with "timestep" set to 1.0.
2. To keep longer timesteps with all bonds to H rigid, "rigidBonds" value is put to all, and "timestep" is set to 2.0.
3. Additional rigid bonds options can be manually adjusted.

3.5.7 Initial Velocities

1. Initial velocities are obtained from Boltzmann distribution by setting "temperature" at 300 or from restart file "velocities restart.vel."
2. Center of mass motion is subtracted.

3.5.8 Number of Steps

1. Based on the required timescale of the simulation model, the number of steps is set.
2. For a continuing simulation, first, the number of steps already done is given as "firstTimestep 30,000" (defaults to 0).
3. The number of steps required for the production run must be specified in the input file. In our example, the run is set to "numsteps 40,000" which must be greater than or equal to the first timestep.

3.5.9 Restart and Trajectory Output

1. "outputName run1" (generates run1.coor, etc.) "restartFreq 1000" (save every 1000 steps) represents the final configuration and the periodic restart files, respectively.
2. Based on the following command "DCDFreq 500," a DCD trajectory file will be generated with 1 frame every 500 steps, and the previous files are renamed to. BAK (*see* **Notes 9–12**).

3.6 Analysis

3.6.1 Structural Comparison of AMPs in Membrane Model

AMP is well known to undergo a structural transition from an unfolded state in solution to a folded conformation within the membrane-embedded system. In this section, we will analyze the key interaction and relevant details supported by MD simulation. The analysis conferred in this chapter relies on a previous study [32].

From the simulation timescale, the analysis reveals that the C-terminal helix of CecB-E53 tends to unfold in cases, where it is not interacting with the phosphate head groups (Figs. 1 and 2). On the other hand, the amphipathic N-terminal helix was found to interact with the negatively charged head groups of the outer layer (phosphate groups) through electrostatic interaction. The positively charged residues include Arg at position 27, and Lys at positions 29, 32, 33, 36, and 39 in CecB-E53 are the main contributing factors in the abovementioned interaction. Substitution of E53 residue with an uncharged residue (Q53) influences the conformational stability of the hydrophobic region (A^{48}GPAIQVLGSAKAI^{61}) in CecB-Q53. Here, inter-residue bonds are formed because of the presence of each H bond donor and acceptor within the Q53 side chain, which aids stabilization to the helix-loop-helix region. This hydrophobic region of peptide acts as the interacting surface to the targeted membrane. The root mean square deviation (RMSD) reveals that the structure of Q53 is stable (converged) in the membrane system. However, in E53, a high RMSD is evident due to the reorganization of the helix-loop-helix structure, which also

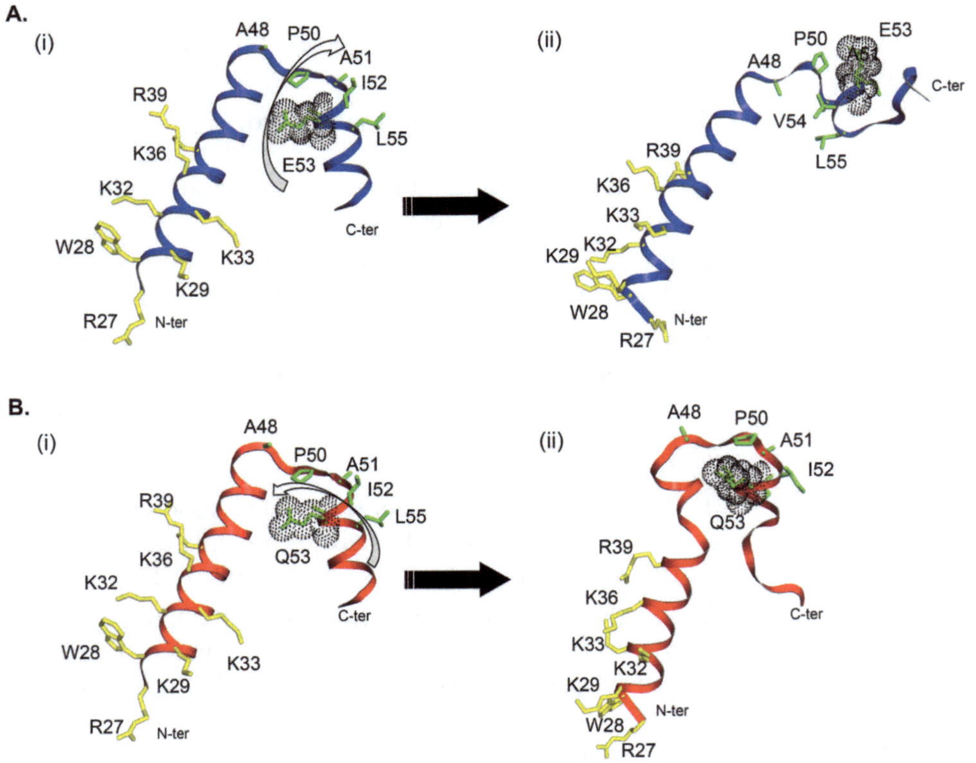

Fig. 1 Conformational dynamics from MD simulation of (**a**) E53 and (**b**) Q53 variants in POPE/POPG (3:1) system. Directionality of conformational changes is shown with arrowhead ((I) 0 ns and (ii) 50 ns). Yellow and green color code denotes residues involved in interaction with membrane and stabilization of hydrophobic loops, respectively. (Figure is adapted with permission from Romoli et al. [32] © 2019. American Chemical Society)

indicates that the C-terminal region, in this case, is dynamic. The substitution of glutamic acid in 53 positions by glutamine changes the overall hydropathy of the peptide [38]. Thus, the weaker antimicrobial activity of CecB-E53 in comparison to CecB-Q53 is attributed to its lower hydrophobicity and the helical nature of CecB-E53 within the lipid environment.

3.6.2 Membrane Perturbation Analysis

The area per lipid tool in VMD membrane plugin [29] analyzes the area per lipid moiety in the membrane architecture. Individual lipid moiety in each leaflet is analyzed in membrane simulation, by denoting "0" and "1" to the upper and lower leaflet, in the analysis. The average value is plotted by choosing each of the leaflets. In our study, 50 frames are evaluated per iteration to calculate the average area per lipid, influenced by the interaction of peptide. The fluctuations in the area per lipid are influenced with the compressibility modulus, K_A, according to the equation:

$$K_{\mathrm{A}} = \frac{2k_B(T)(A_{\mathrm{L}})}{N_{\mathrm{L}}\sigma^2}$$

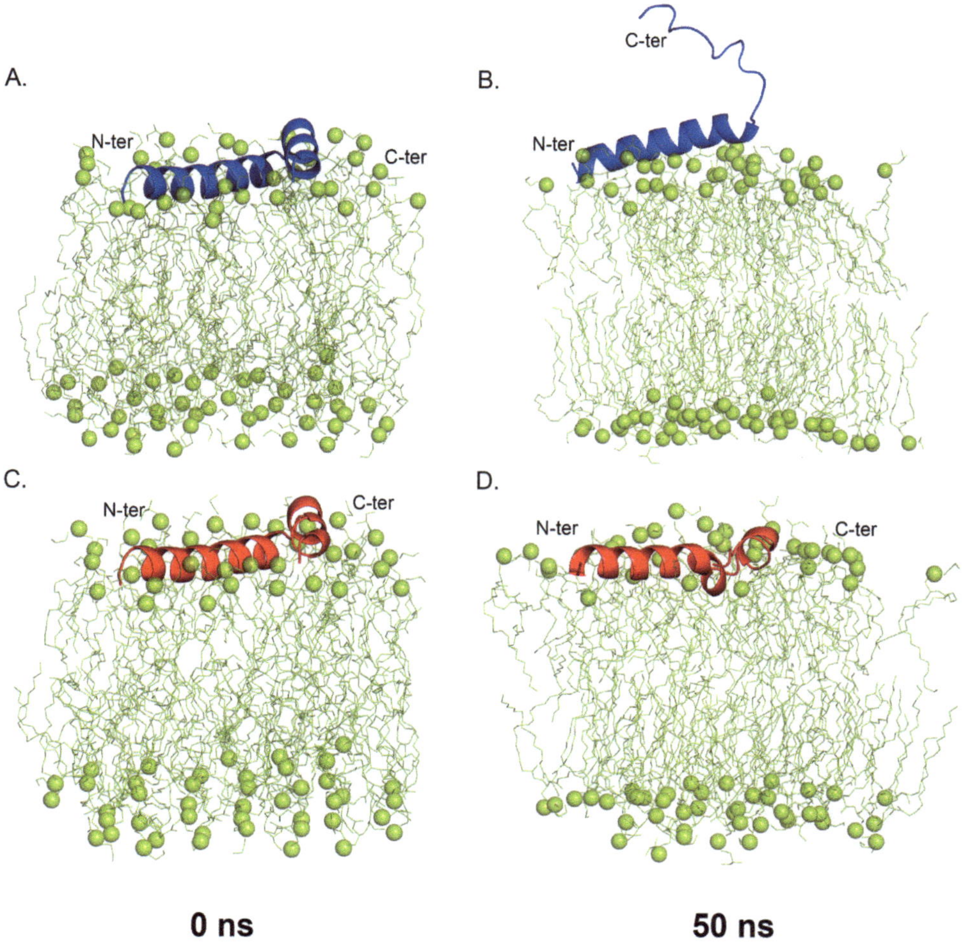

Fig. 2 Models showing the overall dynamicity of E53 (**a** and **b**) and Q53 (**c** and **d**) in POPE/POPG system (3:1) from MD simulation. (Figure adapted with permission from Romoli et al. [32] © 2019. American Chemical Society)

where Boltzmann constant is denoted by k_B, the variance associated with A_L is denoted by σ^2, N_L is the number of lipids, and time and ensemble averages are denoted by the angle brackets.

The variation in the area per lipid shows that membrane permeability increases in the presence of positively charged AMP. From the trajectory analysis, the area per lipid (average) in POPG increases to 63.95 Å and 61.51 Å, respectively, in Q53 and E53 variants. The average of the control membrane system, without the inserted AMP, however, is 59.52 Å [36]. The difference in the area per lipid thus shows that negatively charged residue causes more repulsive perturbation (Fig. 3a, b).

The perturbation induced with the inclusion of peptide into the lipid bilayer is also responsible for the change in direction and versatility of the C–H bond, which can be studied by calculating the lipid order. The value of the order parameter ($-S_{CD}$) is indicative of the membrane fluidity, which can be calculated using the equation:

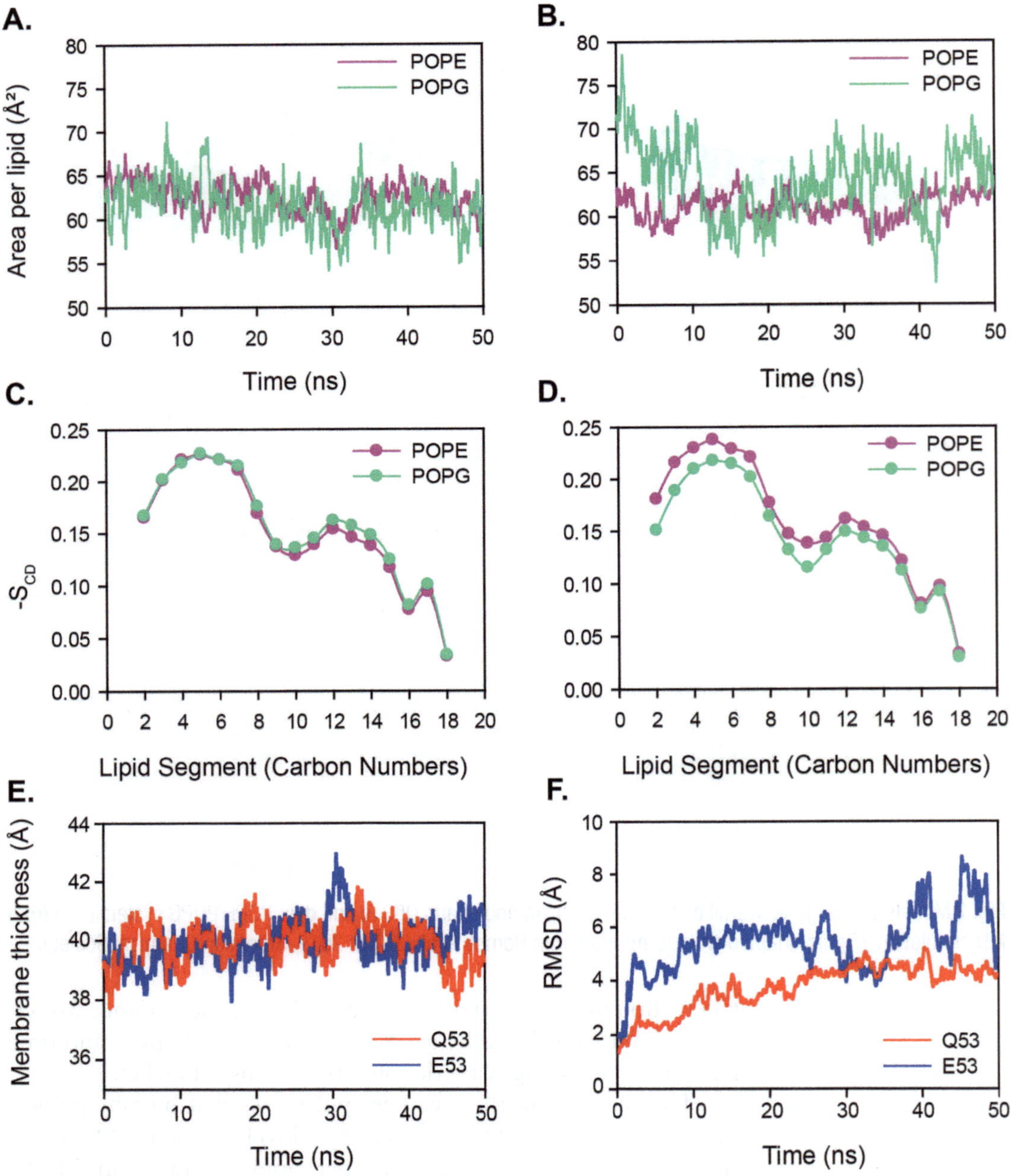

Fig. 3 Area per lipid of POPE/POPG (3:1) system in presence of E53 (**a**) and Q53 (**b**). Lipid order parameter ($-S_{CD}$) of POPE and POPG lipid moieties in presence of E53 (**c**) and Q53 (**d**). (**e**) Average membrane thickness of the model membrane with respect to simulation timescale. (**f**) RMSD of the peptide conformation. The analysis was performed using MEMPLUG. (Figure adapted with permission from Romoli et al. [32] © 2019. American Chemical Society)

$$-S_{\mathrm{CD}} = -\frac{1}{2}\left(3\cos^2 y - 1\right)$$

In this equation, y is the angle between C–H bond and the normal bilayer. To calculate the order parameter for that particular lipid, corresponding atom names (of lipid chains from specific lipid moieties) are required to be selected. Chapman et al. have outlined that membranous architecture in the biological system is majorly heterogeneous and thus show two phases, impacting the permeability and fluidness of the bilayer [39]. Here, the higher $-S_{\mathrm{CD}}$ value denotes increased lipid order. In Fig. 3c and d, both E53 and Q53 showed a decaying pattern, whereas the former AMP is maintaining a more ordered orientation within the lipid bilayer in comparison to the latter. This is indicative of stable interaction between Q53 and anionic phosphate head group of POPG.

The membrane thickness tool calculates the partition between two lipid moieties in upper and lower leaflets (phosphate atoms) with reference to the center of mass. The focal point of lipid particle is standardized relying upon the mass thickness of the specified lipid molecule along the normal of the bilayer. Here, layer thickness refers to the "phosphate-to-phosphate" division, which is influenced by the acyl chain length, the degree of unsaturation of lipid, and the tilt angle. The heterogeneous POPE/POPG (3:1) bilayer shows a normal layer thickness of 41.65 Å, which is reduced to 38.7 Å, and 39.5 Å in E53 and Q53 inserted system. Here, the association between the head groups and cationic residues of AMP might be responsible for the thinning impact, as shown in Fig. 3e. The RMSD plot showing the structural convergence is shown in Fig. 3f.

4 Notes

1. The *PDB Reader* in CHARMM-GUI is the starting point of modeling.
2. PDB coordinates of AMPs contain the structural coordinates. Note that PDB file does not contain the information for peptide orientation with respect to lipid bilayers. *Membrane Builder* tools [40] are useful in this context, to orient the molecule and refer to the pre-oriented protein structures from the OPM database (http://opm.phar.umich.edu). Selecting OPM as a PDB download source is an appropriate method.
3. Membrane Builder supports rectangular and hexagonal framework shapes inside the *XY* plane. The protein inserted lipid bilayer can be made in two different ways: (1) the substitution technique that utilizes lipid-like pseudo-particles and replaces them by lipid atoms (it gets a pleasantly pressed lipid layer around the protein) and (2) inclusion procedure, in which a pre-equili-

brated lipid bilayer is utilized with an opening of a proper size into that a protein is embedded (this strategy is useful for a protein with a tube-shaped and symmetrical structure).

4. The Solvator module solvates the particle or produces a water box for various purposes. A default estimation of 10 Å is utilized for the hole between limits of a biomolecule and edge of solvation box, which might be balanced manually. Solvator creates a default cubiform water box with the longest hub except if generally determined. Ions are included in the solvated framework for balance. The default ion fixation is 0.15 M KCl, which is similar to the physiological concentration. The quantities of particles (N_1 and N_2) are precisely dictated by the particle accessible volume (V) that compares to those concluded from the linearized Poisson-Boltzmann (PB) hypothesis.
5. NAMD package contains the codes required to run MD simulation of peptide–lipid frameworks using the CHARMM36 force fields. However, other suitable force fields can be selected for studying a similar system, but the appropriateness can be found with benchmark studies.
6. The periodic boundary conditions are important to approximate large infinite system and avoid self-interaction of macromolecules. The cutoff boundary is thus crucial in this regard. Is a basic parameter while reproducing a perceptible framework utilizing a set number of particles.
7. van der Waals (vdW) interaction (nonbonded interaction) is a critical driving force that tends to govern the interaction phenomenon in the biomolecular system. In NAMD, the cutoff distance is used to truncate the vdW interactions, specified by keyword *cutoff*. The switching parameter mainly affects the vdW interaction. When this option is set to on, a smooth switching function will be used at the cutoff distance to truncate the vdW potential energy. If switching is set to off, it abruptly truncates the vdW energy at the cutoff distance, so that power is not conserved. The value of *switchdist* should be less than the value of *cutoff*.
8. All improper atom–atom interactions and steric hindrances are removed and relaxed during the minimization of the peptide. In this progression, the peptide is energetically minimized concerning the topology record.
9. RMSD is used as a standardized measure to estimate the least representative of the structural similarity in the conformational hyperspace. RMSD value <2 Å is indicative of structural convergence or stability.
10. Cluster analysis is also another useful method to determine the conformational variability.

11. The electrostatic forces mediate the interaction between the charged peptide and the negatively charged phosphate head groups of lipids. In an unbiased MD simulation, it is encouraged to put the peptide near the lipid bilayer *XY* plane [41] (≤10 Å) to achieve an equilibration state at a moderate timescale.
12. The membrane perturbation induced with insertion of the peptide can be studied by calculating the area per lipid, membrane thickness, and lipid order parameters. These can be computed using the membrane plugin tool.

References

1. Venable RM, Krämer A, Pastor RW (2019) Molecular dynamics simulations of membrane permeability. Chem Rev 119(9):5954–5997
2. Karplus M, McCammon JA (2002) Molecular dynamics simulations of biomolecules. Nat Struct Mol Biol 9(9):646
3. Lee J, Jung SW, Cho AE (2016) Molecular insights into the adsorption mechanism of human β-Defensin-3 on bacterial membranes. Langmuir 32(7):1782–1790. https://doi.org/10.1021/acs.langmuir.5b04113
4. Amos ST, Vermeer LS, Ferguson PM, Kozlowska J, Davy M, Bui TT, Drake AF, Lorenz CD, Mason AJ (2016) Antimicrobial peptide potency is facilitated by greater conformational flexibility when binding to Gram-negative bacterial inner membranes. Sci Rep 6:37639. https://doi.org/10.1038/srep37639
5. Hultmark D, Steiner H, Rasmuson T, Boman HG (1980) Insect immunity. Purification and properties of three inducible bactericidal proteins from hemolymph of immunized pupae of Hyalophora cecropia. Eur J Biochem 106(1):7–16
6. Bulet P, Stocklin R (2005) Insect antimicrobial peptides: structures, properties and gene regulation. Protein Pept Lett 12(1):3–11
7. Lee T-H, N Hall K, Aguilar M-I (2016) Antimicrobial peptide structure and mechanism of action: a focus on the role of membrane structure. Curr Top Med Chem 16(1):25–39
8. Bahar A, Ren D (2013) Antimicrobial peptides. Pharmaceuticals 6(12):1543–1575
9. Kang H-K, Kim C, Seo CH, Park Y (2017) The therapeutic applications of antimicrobial peptides (AMPs): a patent review. J Microbiol 55(1):1–12
10. Radek K, Gallo R (2007) Antimicrobial peptides: natural effectors of the innate immune system. In: Seminars in immunopathology, vol 1. Springer, New York, pp 27–43
11. Kouranov A, Xie L, de la Cruz J, Chen L, Westbrook J, Bourne PE, Berman HM (2006) The RCSB PDB information portal for structural genomics. Nucleic Acids Res 34(suppl_1):D302–D305
12. Nosé S (1984) A molecular dynamics method for simulations in the canonical ensemble. Mol Phys 52(2):255–268
13. Sharma V, Belevich G, Gamiz-Hernandez AP, Róg T, Vattulainen I, Verkhovskaya ML, Wikström M, Hummer G, Kaila VR (2015) Redox-induced activation of the proton pump in the respiratory complex I. Proc Natl Acad Sci U S A 112(37):11571–11576
14. Ulmschneider JP, Ulmschneider MB (2018) Molecular dynamics simulations are redefining our view of peptides interacting with biological membranes. Acc Chem Res 51(5):1106–1116
15. Marrink SJ, De Vries AH, Tieleman DP (2009) Lipids on the move: simulations of membrane pores, domains, stalks and curves. Biochim Biophys Acta 1788(1):149–168
16. Nelson MT, Humphrey W, Gursoy A, Dalke A, Kalé LV, Skeel RD, Schulten K (1996) NAMD: a parallel, object-oriented molecular dynamics program. Int J High Perform Comput Appl 10(4):251–268
17. Vanommeslaeghe K, Hatcher E, Acharya C, Kundu S, Zhong S, Shim J, Darian E, Guvench O, Lopes P, Vorobyov I (2010) CHARMM general force field: a force field for drug-like molecules compatible with the CHARMM all-atom additive biological force fields. J Comput Chem 31(4):671–690
18. Leonard AN, Pastor RW, Klauda JB (2018) Parameterization of the CHARMM all-atom force field for ether lipids and model linear ethers. J Phys Chem B 122(26):6744–6754

19. Kubelka J, Chiu TK, Davies DR, Eaton WA, Hofrichter J (2006) Sub-microsecond protein folding. J Mol Biol 359(3):546–553
20. Voter AF (1997) A method for accelerating the molecular dynamics simulation of infrequent events. J Chem Phys 106(11):4665–4677
21. Bussi G, Laio A, Parrinello M (2006) Equilibrium free energies from nonequilibrium metadynamics. Phys Rev Lett 96(9):090601
22. Grubmüller H (1995) Predicting slow structural transitions in macromolecular systems: conformational flooding. Phys Rev E 52(3):2893
23. Darve E, Pohorille A (2001) Calculating free energies using average force. J Chem Phys 115(20):9169–9183
24. Banerjee V, Kar RK, Datta A, Parthasarathi K, Chatterjee S, Das KP, Bhunia A (2013) Use of a small peptide fragment as an inhibitor of insulin fibrillation process: a study by high and low resolution spectroscopy. PLoS One 8(8):e72318
25. Kar RK, Brender JR, Ghosh A, Bhunia A (2018) Nonproductive binding modes as a prominent feature of Aβ40 Fiber elongation: insights from molecular dynamics simulation. J Chem Inf Model 58(8):1576–1586
26. Schwede T, Kopp J, Guex N, Peitsch MC (2003) SWISS-MODEL: an automated protein homology-modeling server. Nucleic Acids Res 31(13):3381–3385
27. Jo S, Kim T, Iyer VG, Im W (2008) CHARMM-GUI: a web-based graphical user interface for CHARMM. J Comput Chem 29(11):1859–1865. https://doi.org/10.1002/jcc.20945
28. Humphrey W, Dalke A, Schulten K (1996) VMD: visual molecular dynamics. J Mol Graph 14(1):33–38
29. Guixa-González R, Rodriguez-Espigares I, Ramírez-Anguita JM, Carrio-Gaspar P, Martinez-Seara H, Giorgino T, Selent J (2014) MEMBPLUGIN: studying membrane complexity in VMD. Bioinformatics 30(10):1478–1480
30. Lee J, Cheng X, Swails JM, Yeom MS, Eastman PK, Lemkul JA, Wei S, Buckner J, Jeong JC, Qi Y (2015) CHARMM-GUI input generator for NAMD, GROMACS, AMBER, OpenMM, and CHARMM/OpenMM simulations using the CHARMM36 additive force field. J Chem Theory Comput 12(1):405–413
31. Lee E, Jeong K-W, Lee J, Shin A, Kim J-K, Lee J, Lee DG, Kim Y (2013) Structure-activity relationships of cecropin-like peptides and their interactions with phospholipid membrane. BMB Rep 46(5):282
32. Romoli O, Mukherjee S, Mohid SA, Dutta A, Montali A, Franzolin E, Brady D, Zito F, Bergantino E, Rampazzo C (2019) Enhanced silkworm Cecropin B antimicrobial activity against *Pseudomonas aeruginosa* from single amino acid variation. ACS Infect Dis 5(7):1200–1213
33. Roy A, Kucukural A, Zhang Y (2010) I-TASSER: a unified platform for automated protein structure and function prediction. Nat Protoc 5(4):725–738. https://doi.org/10.1038/nprot.2010.5
34. Rohl CA, Strauss CE, Misura KM, Baker D (2004) Protein structure prediction using Rosetta. In: Methods in enzymology, vol 383. Elsevier, Amsterdam, pp 66–93
35. Kelm S, Shi J, Deane CM (2010) MEDELLER: homology-based coordinate generation for membrane proteins. Bioinformatics 26(22):2833–2840
36. Mukherjee S, Kar RK, Nanga RPR, Mroue KH, Ramamoorthy A, Bhunia A (2017) Accelerated molecular dynamics simulation analysis of MSI-594 in a lipid bilayer. Phys Chem Chem Phys 19(29):19289–19299
37. Brooks BR, Brooks IIICL, Mackerell AD Jr, Nilsson L, Petrella RJ, Roux B, Won Y, Archontis G, Bartels C, Boresch S (2009) CHARMM: the biomolecular simulation program. J Comput Chem 30(10):1545–1614
38. Chen Y, Guarnieri MT, Vasil AI, Vasil ML, Mant CT, Hodges RS (2007) Role of peptide hydrophobicity in the mechanism of action of alpha-helical antimicrobial peptides. Antimicrob Agents Chemother 51(4):1398–1406. https://doi.org/10.1128/AAC.00925-06
39. Chapman D, Urbina J, Keough KM (1974) Biomembrane phase transitions studies of lipid-water systems using differential scanning calorimetry. J Biol Chem 249(8):2512–2521
40. Wu EL, Cheng X, Jo S, Rui H, Song KC, Dávila-Contreras EM, Qi Y, Lee J, Monje-Galvan V, Venable RM (2014) CHARMM-GUI membrane builder toward realistic biological membrane simulations. J Comput Chem 35(27):1997–2004
41. Mihailescu D, Smith JC (2000) Atomic detail peptide-membrane interactions: molecular dynamics simulation of gramicidin S in a DMPC bilayer. Biophys J 79(4):1718–1730

Part V

Insect Immunity and Environmental Interactions

Chapter 11

Manipulating the Mosquito Microbiota to Study Its Function

Ottavia Romoli and Mathilde Gendrin

Abstract

Aedes aegypti mosquitoes are the main vectors of several arboviruses and are commonly used as models in mosquito biology and vector competence studies. The mosquito microbiota has an impact on different aspects of host physiology, including development, immunity, and fecundity, in turn influencing the capability of the mosquito to transmit diseases. The composition of the microbiota is relatively simple in field mosquitoes, and many of its bacterial members are culturable in the laboratory. Being able to manipulate the composition of the mosquito microbiota is essential to effectively investigate its effect on host physiology and vector competence. This protocol describes how to obtain gnotobiotic mosquitoes, i.e., mosquitoes with a known microbiota composition, and how to monitor the effect of a manipulated microbiota on mosquito development.

Key words Mosquito, Microbiota, *Aedes aegypti*, Development, Gnotobiology, Wing length

1 Introduction

The mosquito *Aedes aegypti* is the main vector of several arboviruses, including yellow fever, chikungunya, Zika, and dengue viruses. The distribution of this mosquito species throughout all tropical and subtropical regions of the world and the ease of rearing it in the laboratory makes it a commonly used model for mosquito biology and vector competence studies.

The mosquito microbiota is known to have a significant impact on several factors determining disease transmission, including mosquito vector competence (i.e., capability to acquire and transmit pathogens) and different aspects of mosquito physiology. The larval microbiota has a critical effect on host physiology, since larval development is only possible in the presence of a living microbiota [1]. The rescue of development in axenic (i.e., germ-free) larvae can be achieved with specific diet supplementations, but with significant delay in larval growth [2]. Exposure to different

Federica Sandrelli and Gianluca Tettamanti (eds.), *Immunity in Insects*, Springer Protocols Handbooks,
https://doi.org/10.1007/978-1-0716-0259-1_11,

bacteria during larval development leads to significant differences in developmental rate, adult body size, and egg production [3, 4]. The adult microbiota composition also influences host physiological traits such as lifespan, mating behavior, and vector competence [5, 6].

The mosquito microbiota has a relatively low bacterial diversity, with four bacterial taxa representing on average 90% of the microbial population in the gut of a single mosquito [7]. Both culture-dependent and culture-independent analyses showed that the mosquito microbiota is mainly composed of Gram-negative bacteria, of which many have been cultured [7–10]. Due to its culturability and simple composition, the mosquito microbiota can be manipulated in the laboratory, and this represents a useful tool to study the role of the microbiota on mosquito physiology.

Here we describe a method to obtain gnotobiotic *Aedes aegypti* mosquitoes, i.e., mosquitoes that are colonized by a microbiota of known composition, in the present case composed of a single bacterial strain. In conventional rearing conditions, microbes are transmitted horizontally, between individuals sharing the same environment, or vertically, mostly via contamination of the surface of mosquito eggs by a female during egg laying. The egg cytoplasm is generally microbiologically sterile. When larvae hatch, they acquire their microbiota through ingestion of microorganisms present in the rearing water and/or on the egg envelope. Using this gnotobiology protocol, axenic *Aedes aegypti* larvae are obtained by surface sterilization of mosquito eggs (Fig. 1). After addition of a bacterial culture to the rearing water, gnotobiotic larvae and mosquitoes are obtained and can be used to study the impact of a manipulated microbiota on mosquito development. Two different rearing methods are used depending on the parameters that need to be measured. Rearing larvae in a 24-well plate allows the follow-up of single individuals throughout

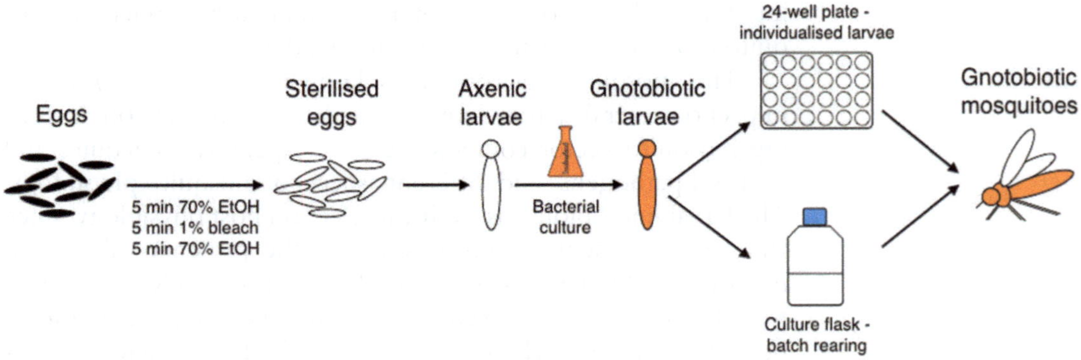

Fig. 1 Experimental setup to obtain gnotobiotic mosquitoes. Eggs are surface sterilized with three subsequent 5-min washes in 70% ethanol, 1% bleach, and 70% ethanol and rinsed in sterile water. A bacterial culture is added to axenic larvae to obtain gnotobiotic larvae and, subsequently, gnotobiotic adult mosquitoes. Gnotobiotic larvae can be reared in 24-well plates, if parameters on single individuals have to be monitored, or in cell culture flasks

Table 1
Experimental timeline of the gnotobiology protocol

Day	Protocol step	
0	Revive the bacterial strain (Subheading 3, **step 1**)	
1	Sterilize *Ae. aegypti* eggs (Subheading **3**, **step 2**) and inoculate the bacterium (Subheading **3**, **step 3**)	
2	Generate gnotobiotic larvae (Subheading **3**, **step 4**)	
	24-well plates	**Culture flasks**
3	Determine duration of developmental stages (Subheading **3**, **step 6c**)	
4		
5		
6		
7	Measure larval length (Subheading **3**, step **6b**)	Measure larval length (Subheading **3**, **step 6b**)
8	Determine developmental success/sex ratio (Subheading **3**, **step 6e**)	Transfer pupae to sterile boxes (Subheading **3**, **step 5**)
9		
10		
11	Start survival experiments (Subheading **3**, **step 6f**)	
12		
13	Collect individuals for wing length determination (Subheading **3**, **step 6d**)	

development, while batch rearing in cell culture flasks is more time-efficient and therefore preferable when the measured parameters allow grouping. The experimental timeline of the protocol is indicated in Table 1.

Besides the rearing methodology, we provide an overview of different practices that are commonly used to estimate larval and mosquito development: larval length, duration of developmental stages, development success rate, adult sex ratio, adult wing length, and mosquito lifespan. Larval length is a parameter that is strictly dependent on larval rearing conditions. Notably, it varies significantly depending on larval density [11], rearing temperature [12], and diet [13]. In addition to larval length, larval development can also be quantified in terms of duration of developmental stages and of developmental success rate, which both depend on nutrition and microbiota [3, 14]. The sex ratio of emerged adults is also influenced by larval rearing conditions [11]. In *Aedes aegypti*, sex is completely genetically determined [15] and males have a faster larval development than females. Therefore, although the reasons behind why rearing conditions affect sex ratio have not been determined, one can theorize that the presence of a larval slow killer could result in a higher proportion of adult males. Wing length is commonly used as a read-out of adult body size, which is dependent on larval life-history traits. The adult body size is an important parameter, as it is known to directly correlate with blood meal volume and fecundity [16] and has been reported to be inversely correlated with dengue virus dissemination in the mosquito [17]. Finally, mosquito lifespan is a good indicator of the effect of the microbiota on the adult physiology. Considering that the transmitted pathogens develop for 1–2 weeks in the mosquito before being infectious for humans, mosquito lifespan is an important determinant of transmission.

Depending on the microbe and mosquito species used to generate gnotobiotic mosquitoes, this protocol should be adapted for the chosen host-microbe couple. In particular, culture media and microbial concentrations may vary from one microorganism to another, and parameters of egg sterilization may vary between mosquito species. Here, we describe the materials and the procedures that we use to work with *Escherichia coli* in *Ae. aegypti*. However, several microbial strains have been used to produce gnotobiotic larvae of *Ae. aegypti* and *Anopheles gambiae* in independent laboratories [1, 3, 14].

2 Materials

2.1 Live Organisms

1. Glycerol *E. coli* stock kept at −80 °C.
2. *Ae. aegypti* eggs, kept on a dry paper for 4 days to 3 months (*see* **Note 1**).

2.2 Media/Solutions

1. Liquid Luria-Bertani (LB) medium: 10 g/L tryptone, 5 g/L yeast extract, 10 g/L NaCl*.
2. Solid LB medium: 10 g/L tryptone, 5 g/L yeast extract, 10 g/L NaCl, 1.5 g/L agar*.
3. Sterile deionized water*.
4. 70% ethanol, prepared with sterile deionized water.
5. 1% bleach, freshly prepared from a tablet into sterile deionized water (bleach concentration refers to the active chlorine).
6. Sterile 5% (w/v) fish food (TetraMin Baby) in deionized water (*see* **Note 2**)*.
7. 4% paraformaldehyde in autoclaved phosphate-buffered saline*: PBS, 8 g/L NaCl, 0.2 g/L KCl, 1.44 g/L Na_2HPO_4, 0.24 g/L KH_2PO_4. Add paraformaldehyde just before use (manipulate concentrated paraformaldehyde under a fume hood).
8. Sterile 10% (w/v) sucrose solution*.

*Autoclaved at 121 °C for 30 min.

2.3 Disposable Materials

1. Disposable vacuum filtration unit (the material and the pore size of the filtering membrane are not important; we generally use nitrocellulose 0.2 μm membranes).
2. Sterile 25-mL cell culture flasks with vented caps.
3. Sterile petri dishes.
4. Sterile 24-well plates.
5. Sterile plastic pipettes.
6. Sterile 15-mL tubes.
7. Microscope slides.
8. Coverslips.
9. Autoclavable polypropylene boxes used for plant culture, equipped with a larger tube (typically used for urine collection) for adult emergence and a smaller tube with a cotton roll for mosquito sugar feeding. Cover the filter on the top of the lid with gauze to avoid mosquito contamination. *See* Fig. 2 for more details.

2.4 Other Materials

1. Microbiological safety cabinet (MSC).
2. Pump for vacuum filtration.
3. Dissection forceps.
4. Dissection microscope with camera.

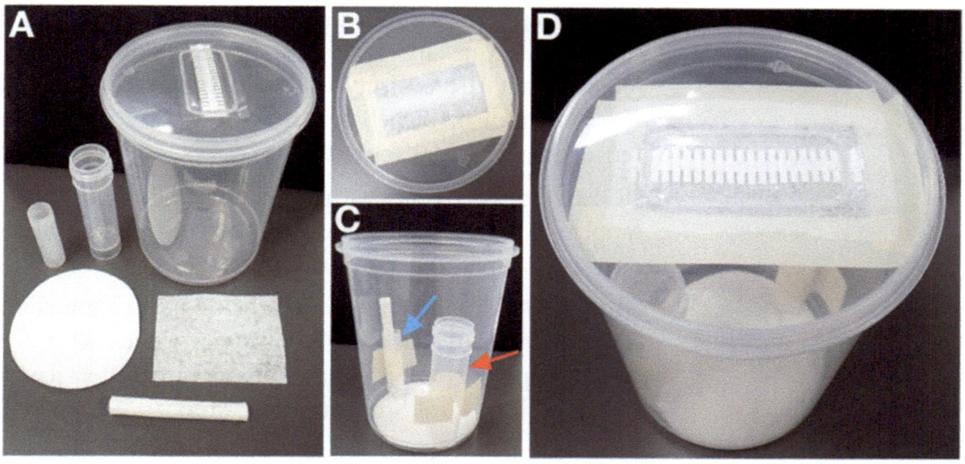

Fig. 2 Preparation of sterile boxes for gnotobiotic adult mosquitoes. (**a**) Material needed to set up the boxes: autoclavable polypropylene box for plant culture, larger autoclavable tube (typically used for urine collection), smaller autoclavable tube, cotton roll, gauze, filter paper cut to fit the bottom of the box. (**b**) The gauze is taped to cover the filter on the internal part of the lid to avoid mosquitoes being exposed to potential contaminants via their proboscis. (**c**) The filter paper is fixed to the bottom of the box with adhesive tape to collect mosquito excreta. The two tubes are also taped to the box sides: the larger one (red arrow) will allocate pupae, while in the smaller one (blue arrow), a cotton roll is placed for sugar feeding. (**d**) The prepared box is ready to be autoclaved: close the lid for approximately 3/4 of the surface, leaving one border of the lid opened. Close the lid immediately after sterilization. Alternatively, boxes can be autoclaved in autoclavable bags. After autoclave, let boxes dry inside a MSC. Before use, add the sterile 10% sucrose solution to the smaller tube with the cotton roll

3 Methods

1. The day before egg sterilization (*day 0*), revive the *E. coli* strain from the glycerol stock by streaking it onto an LB agar plate. Incubate the plate overnight at 37 °C.
2. *Sterilization of Ae. aegypti eggs—day 1* (this procedure should be conducted inside a MSC):
   (a) Transfer *Ae. aegypti* eggs onto the filter of a vacuum filtration apparatus system. Since the sterilization process can reduce the percentage of hatched eggs, we suggest exceeding the amount of eggs with respect to the required number of larvae by 10–15%. Note that the filtration unit does not need to be sterile as it is only used to remove sterilizing solutions. The same filtration unit can be used several times, as long as the filter is not visibly damaged.
   (b) From this point, proceed with egg sterilization without stopping between steps to avoid egg desiccation.
   (c) Add ~50 mL of 70% ethanol and incubate 5 min. During this incubation, mix the solution with a P1000 pipette tip to ensure that all eggs are well submerged in ethanol (*see* **Note 3**). Apply the vacuum to remove the ethanol.
   (d) Repeat the previous step with ~50 mL of 1% bleach (instead of ethanol) for 5 min and then again with ~50 mL of 70% ethanol for another 5 min. The volumes of sterilizing solutions are approximate: the critical step is to ensure that the entirety of the egg surface is in contact with ethanol or bleach for the correct amount of time to ensure bactericidal activity.
   (e) Rinse eggs three times with abundant sterile deionized water, applying the vacuum each time to remove the water. This step is essential to remove any traces of sterilizing solutions.
   (f) Add sterile deionized water to the eggs, and transfer them to a cell culture flask for larval emergence using a sterile plastic pipette. Keep the hatching larvae in their flask without any food until Subheading 3, **step 4**.
   (g) To check the efficiency of egg surface sterilization, transfer 10–20 eggs to a sterile 15-mL conical tube containing 3 mL of LB, and incubate shaking at 37 °C overnight.
3. Inoculate a single *E. coli* colony into LB, and incubate shaking at 37 °C overnight (for ~16 h, *see* **Notes 4** and **5**)—*day 1*.
4. *Generation of gnotobiotic Ae. aegypti larvae—day 2*:
   (a) Centrifuge the *E. coli* culture for 30 min at 3200 × *g*. Resuspend the pellet in the same volume of sterile deionized water. Dilute this suspension 1:5, and check that the

resulting suspension contains approximately 10^8 CFU/mL by plating on LB agar. This bacterial suspension will be added to larvae (*see* **Note 4**).

(b) *24-well plate setup*: transfer a single axenic larva to each well using a sterile plastic pipette. Add 2 mL of the bacterial suspension and 50 μL of the sterile diet suspension to each well.

(c) *Flask setup*: transfer 10 to 15 axenic larvae to each flask using a sterile plastic pipette. Add 15 mL of the bacterial suspension and 750 μL of the sterile diet suspension.

(d) Keep an aliquot of axenic larvae in sterile water in a flask for a few days to ensure that larvae were not contaminated during manipulation. Axenic larvae are not able to grow without bacteria: these larvae will molt to the second instar (whether food was added or not) only if contamination has occurred during the plate/flask setup.

(e) Ensure that plates and/or flasks are closed to maintain sterility. Place larvae in a climatic chamber (30 °C with LD 12:12 h cycle) or in an insectary compartment specifically dedicated to microbiota-manipulated mosquitoes.

5. *Maintaining gnotobiotic Ae. aegypti mosquitoes—days 8–10.* When the pupal stage is reached, transfer pupae to the autoclaved polypropylene boxes (Fig. 2). Pupae are placed in the larger tube, while a sterile 10% sucrose solution is added to the small tube containing the sterile cotton roll.

6. *Monitoring the effect of the manipulated microbiota on mosquito development and survival*:

(a) As a control, use eggs from the same batch that were not subjected to the sterilization process and hatched in non-sterile water. Alternatively, use gnotobiotic eggs from the same batch and from the same sterilization process that are colonized with a control microbe. The rearing conditions of non-sterile larvae should be as similar as possible to those used for the gnotobiotic rearing, using the same flasks or plates and the same sterilized diet. However, non-sterile controls should not be handled in microbiologically sterile conditions, i.e., on the bench rather than inside a MSC.

(b) *Larval length*: 5 days post bacterial inoculation (*day 7*), larvae should have reached the fourth instar, and body length can be used as a proxy for larval development (*see* **Note 6**). For this analysis, larvae can be reared either on 24-well plates or on flasks. Fix larvae for 20 min at room temperature in PBS with 4% paraformaldehyde. Using a clean plastic pipette, place the desired number of larvae on a microscope slide, and remove as much liquid as possible. Several larvae can be placed on the same slide, using clean forceps. Take a picture rapidly to avoid larval desiccation (*see* **Note 7**). The larval length is considered the distance between the anterior border of the head and the posterior border of the last abdominal segment, excluding the siphon (Fig. 3a).

(c) Duration *of developmental stages*: larvae placed in single wells of a 24-well plate are inspected daily, three times per day. For each larva, the moment at which they molt is noted. In this way the duration of each larval instar and of the pupal stage can be determined for single individuals. A good indicator of larval molting is the presence and the size of exuviae in the well; therefore the larval stage can be deduced by counting the number of exuviae (one exuvia = second instar, two exuviae = third instar, three exuviae = fourth instar). Since the first exuvia is often difficult to see, second instar larvae can also be distinguished from first instar larvae by their darker head.

(d) *Wing length*: collect adult mosquitoes 2 days post emergence, as younger mosquitoes may not have completely spread or dried wings, while older mosquitoes might start to have damaged wings. Both rearing conditions, whether in flasks or in plates, are relevant for this analysis. For convenience, adult mosquitoes can be kept at −20 °C until analysis. The measurement can be conducted on both wings or on one specific wing. It is however important to discriminate between male and female mosquitoes because of the significant difference in their wing sizes (Fig. 3b, c). Using dissection forceps, collect wings from several individuals and place them on a microscope slide. Use a coverslip to flatten samples and take a picture using the microscope camera. Wing length is defined as the distance between the alular notch and the radius 3 vein (Fig. 3b, c).

(e) *Percentage of developed larvae and sex ratio*: larvae reared in 24-well plates are left in the plates until the adult stage. For each individual, mark the sex of the adult mosquito or if larvae/pupae are undeveloped or dead. In our hands, gnotobiotic larvae generally reach the adult stage at days 8–10 post bacterial inoculation. Since *Ae. aegypti* larvae might survive several days without developing, we set the end point of the experiment at day 15 post bacterial inoculation, i.e., at least 5 days after observing the last pupation. This analysis allows operators to determine if the gnotobiotic rearing impacts the overall development of mosquitoes (*see* **Note 8**).

(f) *Survival*: transfer pupae in autoclaved polypropylene boxes and provide sugar. Do not exceed 30 pupae per box. Inspect daily the boxes to count the number of dead and possibly determine their sex.

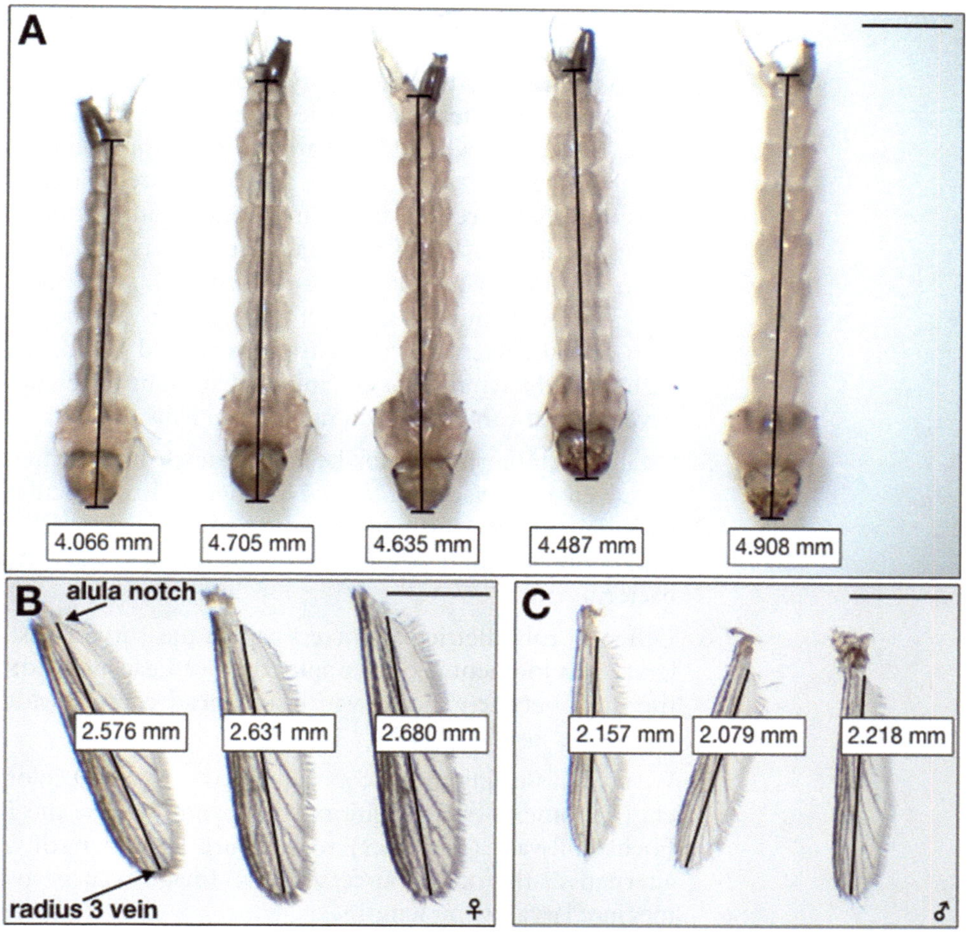

Fig. 3 Measuring larval and wing lengths as a proxy for development. (**a**) Measurement of larval length of 5-day-old *Ae. aegypti* gnotobiotic larvae. Larval length is measured from the anterior border of the head to the posterior border of the last abdominal segment, excluding the siphon. (**b**, **c**) Measurement of wing length on female (**a**) and male (**b**) *Ae. aegypti* gnotobiotic mosquitoes. Wing length is defined as the distance between the alular notch and the radius 3 vein. Bars, 1 mm

4 Notes

1. We use the New Orleans strain, but other *Ae. aegypti* strains have been used to produce gnotobiotic mosquitoes [3, 4].
2. We use TetraMin Baby but we do not exclude that other types of diet can be used. Since TetraMin Baby is insoluble, after autoclaving the diet solution, we prepare 5 mL aliquots that are easier to resuspend. This provides a standardized amount of diet to larvae.
3. The incubation time with ethanol and bleach should not exceed 5 min: a longer treatment will result in a decrease of the percentage of hatched eggs.

4. The volumes of the bacterial culture and of the diluted bacterial suspension should be adapted to the number of gnotobiotic larvae required and to the type of larval rearing: consider that 50 mL of bacterial suspension are needed for each 24-well plate (24 larvae), while 15 mL are sufficient for each flask (10–15 larvae). For *E. coli* we use a fivefold diluted bacterial suspension; therefore we inoculate bacteria in a LB volume corresponding to 1:5 of the final bacterial suspension volume (e.g., 200 mL of LB for 1 L of diluted bacterial suspension). Use a pre-inoculum if the bacterial culture volume is larger than 50 mL. Bacterial concentrations should be adjusted for each type of bacterium as the minimal amount to rescue larval development varies between microorganisms [14].
5. Ideally bacteria should not be in the death phase when they are given to larvae. For *E. coli* we decided to inoculate the bacterial culture 16 h before adding bacteria to sterile larvae. This time schedule should be adjusted for each type of bacterium.
6. Other morphometric parameters can be measured to estimate larval development. For example, the head capsule width and the ratio between thorax width and head capsule width are commonly used [18, 19].
7. We use a dissecting microscope equipped with a HD color camera (Euromex). After acquiring the image, we use the Image Focus software (Euromex) to measure larval length. As an alternative, the open-source software ImageJ can be used to measure larval/wing lengths.
8. To determine if a real effect on sex ratio is present, you may determine the sex of undeveloped larvae and pupae. This can be achieved by PCR using the primers for the male-determining *Nix* gene described in [15].

Acknowledgments

We thank Verena Kircher and Charles Bobin for preliminary work on protocols, Jean-Géraud Issaly for egg production, Guillaume Lacour for training on wing length measurements, Siegfried Hapfelmeier for sharing of bacterial strain, and Emma Wise for English proofreading of the manuscript. This work is funded by the French Government's Investissement d'Avenir program, Laboratoire d'Excellence "Integrative Biology of Emerging Infectious Diseases" (grant no. ANR-10-LABX-62-IBEID), and by ANR JCJC Mosmi to MG (grant no. <ANR-18-CE15-0007>).

References

1. Coon KL, Vogel KJ, Brown MR, Strand MR (2014) Mosquitoes rely on their gut microbiota for development. Mol Ecol 23(11):2727–2739. https://doi.org/10.1111/mec.12771
2. Correa MA, Matusovsky B, Brackney DE, Steven B (2018) Generation of axenic *Aedes aegypti* demonstrate live bacteria are not required for mosquito development. Nat Commun 9(1):4464. https://doi.org/10.1038/s41467-018-07014-2
3. Dickson LB, Jiolle D, Minard G, Moltini-Conclois I, Volant S, Ghozlane A, Bouchier C, Ayala D, Paupy C, Valiente Moro C, Lambrechts L (2017) Carryover effects of larval exposure to different environmental bacteria drive adult trait variation in a mosquito vector. Sci Adv 3(8):e1700585. https://doi.org/10.1126/sciadv.1700585
4. Coon KL, Brown MR, Strand MR (2016) Gut bacteria differentially affect egg production in the anautogenous mosquito *Aedes aegypti* and facultatively autogenous mosquito *Aedes atropalpus* (Diptera: Culicidae). Parasit Vectors 9:375. https://doi.org/10.1186/s13071-016-1660-9
5. Cirimotich CM, Dong Y, Clayton AM, Sandiford SL, Souza-Neto JA, Mulenga M, Dimopoulos G (2011) Natural microbe-mediated refractoriness to *Plasmodium* infection in *Anopheles gambiae*. Science 332:855–858. https://doi.org/10.1126/science.1201618
6. Pike A, Dong Y, Dizaji NB, Gacita A, Mongodin EF, Dimopoulos G (2017) Changes in the microbiota cause genetically modified *Anopheles* to spread in a population. Science 357(6358):1396–1399. https://doi.org/10.1126/science.aak9691
7. Osei-Poku J, Mbogo CM, Palmer WJ, Jiggins FM (2012) Deep sequencing reveals extensive variation in the gut microbiota of wild mosquitoes from Kenya. Mol Ecol 21:5138–5150. https://doi.org/10.1111/j.1365-294X.2012.05759.x
8. Ramirez JL, Souza-Neto J, Cosme RT, Rovira J, Ortiz A, Pascale JM, Dimopoulos G (2012) Reciprocal tripartite interactions between the *Aedes aegypti* midgut microbiota, innate immune system and dengue virus influences vector competence. PLoS Negl Trop Dis 6(3):e1561. https://doi.org/10.1371/journal.pntd.0001561
9. Gendrin M, Christophides GK (2013) The Anopheles mosquito microbiota and their impact on pathogen transmission. In: Manguin S (ed) Anopheles mosquitoes—new insights into malaria vectors. IntechOpen, London. https://doi.org/10.5772/55107
10. Charan SS, Pawar KD, Gavhale CV, Tikhe CV, Charan NS, Angel B, Joshi V, Patole MS, Shouche YS (2016) Comparative analysis of midgut bacterial communities in three aedine mosquito species from dengue-endemic and non-endemic areas of Rajasthan, India. Med Vet Entomol 30:264–277. https://doi.org/10.1111/mve.12173
11. Jannat KN, Roitberg BD (2013) Effects of larval density and feeding rates on larval life history traits in *Anopheles gambiae* s.s. (Diptera: Culicidae). J Vector Ecol 38(1):120–126. https://doi.org/10.1111/j.1948-7134.2013.12017.x
12. Christiansen-Jucht C, Parham PE, Saddler A, Koella JC, Basáñez MG (2014) Temperature during larval development and adult maintenance influences the survival of *Anopheles gambiae* s.s. Parasit Vectors 7:489. https://doi.org/10.1186/s13071-014-0489-3
13. Linenberg I, Christophides GK, Gendrin M (2016) Larval diet affects mosquito development and permissiveness to *Plasmodium* infection. Sci Rep 6:38230. https://doi.org/10.1038/srep38230
14. Valzania L, Martinson VG, Harrison RE, Boyd BM, Coon KL, Brown MR, Strand MR (2018) Both living bacteria and eukaryotes in the mosquito gut promote growth of larvae. PLoS Negl Trop Dis 12(7):e0006638. https://doi.org/10.1371/journal.pntd.0006638
15. Hall AB, Basu S, Jiang X, Qi Y, Timoshevskiy VA, Biedler JK, Sharakhova MV, Elahi R, Anderson MA, Chen XG, Sharakhov IV, Adelman ZN, Tu Z (2015) A male-determining factor in the mosquito *Aedes aegypti*. Science 348(6240):1268–1270. https://doi.org/10.1126/science.aaa2850
16. Clements AN (1992) The biology of mosquitoes, vol 1. Chapman and Hall, London
17. Alto BW, Reiskind MH, Lounibos LP (2008) Size alters susceptibility of vectors to dengue virus infection and dissemination. Am J Trop Med Hyg 79(5):688–695
18. Vogel KJ, Valzania L, Coon KL, Brown MR, Strand MR (2017) Transcriptome sequencing reveals large-scale changes in axenic *Aedes aegypti* larvae. PLoS Negl Trop Dis 11(1):e0005273. https://doi.org/10.1371/journal.pntd.0005273
19. Timmermann SE, Briegel H (1999) Larval growth and biosynthesis of reserves in mosquitoes. J Insect Physiol 45(5):461–470. https://doi.org/10.1016/S0022-1910(98)00147-4

Chapter 12

Discovery and Analysis of RNA Viruses in Insects

Lumi Viljakainen and Jaana Jurvansuu

Abstract

Viruses are the most abundant pathogens on Earth infecting all cellular life forms. Only in recent years have we started to gain knowledge on insect viromes, thanks to the development of sequencing technologies. The discovery and characterization of insect viruses is important for understanding insect-virus interactions, coevolution, and insect immune defenses. We describe here a bioinformatic pipeline for the discovery of RNA viruses from insects based on RNA sequence data and the analysis of insect antiviral immune response against the discovered viruses by using small RNA sequence data.

Key words Virus discovery, RNA viruses, Insect viruses, Bioinformatic pipeline, RNA-seq, sRNA-seq, Small RNAs

1 Introduction

Insects hold the record for the most diverse animal taxon on the Earth [1], and in the recent years, with the aid of high-throughput sequencing methods, we have started to appreciate the rich RNA virome they harbor [2–4]. Research on insect RNA viruses has an important value not only for increasing knowledge on virus biodiversity in the most numerous animal species but also to understand the role of insect microbiome in their biology. For example, in transcriptomic studies addressing questions related to insect immunity, it is noteworthy that the presence of RNA viruses may have an effect on inferences based on gene expression profiles [5–7].

We describe here the bioinformatics process of discovering replicating RNA viruses in insects based on two types of RNA sequence data: RNA-sequence (RNA-seq) and small RNA sequence (sRNA-seq) data. RNA-seq reads are typically 100–150 nucleotides (nt) in length and optimal for the assembly of complete virus genomes. sRNA-seq reads are in maximum 50 nt in length and optimal for identifying viruses against which the host has raised an

Federica Sandrelli and Gianluca Tettamanti (eds.), *Immunity in Insects*, Springer Protocols Handbooks,
https://doi.org/10.1007/978-1-0716-0259-1_12,

RNA interference response. When these short reads are mapped against the virus genomes, those viruses that are attacked by the host RNAi machinery will have a sRNA size distribution peaking at 21–22 nt indicative that the double-stranded RNA has been processed by the host Dicer enzyme.

It is noteworthy that the virome discovered with this process is dependent on the infection phase: non-replicating, latent viruses may remain undiscovered. Furthermore, different regions of the virus genome may be represented by varying amounts of sequencing reads due to differences in transcription levels of ORFs. This may lead to difficulties in the assembly of complete viral genomes. Any assembled virus genome represents a consensus sequence of many genotypes since RNA viruses have high mutation rates [8] and evolve as a quasispecies [9]. Finally, sequences of DNA and retroviruses can be found from RNA-sequencing data [4]. However, constructing a complete DNA virus sequence from RNA-sequencing data is difficult as DNA viruses are often large and not the whole genome is transcribed.

It is assumed that you already have RNA-seq and/or sRNA-seq data in fastq format on hand. You can find guidelines for experimental design to obtain such data in, e.g., [10]. The bioinformatic pipeline for the discovery and analysis of RNA viruses using RNA-seq data involves assembly of the sequencing reads to contigs, identification of virus-derived contigs using similarity search against a virus database, annotation of the virus genomes, quantification of the amount of reads derived from each virus, and phylogenetic analysis of the discovered viruses. In addition, if sRNA-seq data is available, it can be used for determining against which of the viruses the host has an active immune response. You need to have the basic knowledge of Unix/Linux operating system to install the required software.

2 Materials

To be able to run the bioinformatic pipeline outlined below, you need a computer with Unix/Linux operating system. The step that uses the most computing power is the assembly of reads to contigs, which, for example in Trinity, requires approximately 1GB of RAM per one million pairs of Illumina reads. The example command lines we present here do not take into account the possible sample-specific properties of your data, which might have to be addressed, for example, by additional or alternative parameters—you should always read the most recent documentation of each software you use.

3 Methods

3.1 Processing and Analysis of RNA-Seq Data

1. The RNA-seq protocol here assumes you have strand-specific paired-end sequence data generated by the Illumina technology with read length of 100–150 nt. Make sure that your data consists of reads for which adapter sequences have been removed, low-quality nucleotides have been trimmed from both ends of the reads, and low-quality reads have been discarded by using, e.g., Trimmomatic [11] for cleaning the data:

```
trimmomatic PE -threads 8 -phred33 left_reads.
fq.gz right_reads.fq.gz left_reads_paired.fq.gz
left_reads_unpaired.fq.gz right_reads_paired.fq.gz
right_reads_unpaired.fq.gz ILLUMINACLIP:TruSeq3-PE.
fa:2:30:10 LEADING:3 TRAILING:3 SLIDINGWINDOW:4:15
MINLEN:36
```

2. Check the quality and length distribution of the reads before and after trimming using FastQC [12].
3. Assemble the reads to contigs using Trinity [13]:

```
Trinity --seqType fq --max_memory 380G --left left_reads_
paired.fq --right right_reads_paired.fq --CPU 16 --output trin-
ity_run_out
```

4. It is advisable to filter out contigs by a minimum length of e.g., 1000 nt, as the smallest RNA virus genomes are approximately 2000 nt [14]. In this way, the manual work required for subsequent steps is reduced, while allowing some of the virus genomes to be partial:

```
bioawk -c fastx '{ if(length($seq) >= 1000) { print
">"$name; print $seq }}' Trinity.fasta > 1000_
Trinity.fasta
```

5. Carry out a BLASTX similarity search with the contigs against the RefSeq release of viral protein sequences available at the National Center for Biotechnology Information (NCBI [15]). It is advisable to download the viral sequences and use BLAST+, a stand-alone version of NCBI BLAST, on your local computer or a server. Download the virus protein sequences and name the file as virusproteins.fas and run BLASTX search. Here is an example command line:

```
blastx -query 1000_Trinity.fasta -db virusproteins.fas -outfmt
"6 qseqid sseqid stitle qlen slen length evalue qcovs pident
qstart qend" -evalue 1e-10 -culling_limit 1 -out Blastx_1000_
trinity_out.txt
```

   An example of the BLASTX results is shown in Table 1.

6. In our experience, some of the Trinity-assembled virus contigs are derived from the same virus genome and may differ at only one or a few nucleotide positions—A very likely scenario given

Table 1
Example of BLASTX search results for two Trinity contigs. The first contig matches several regions of Lasius niger virus 1, and the second contig matches a small region of Hubei picorna-like virus 59

<table>
<tr><th>qseqid</th><th>sseqid</th><th>stitle</th><th>qlen</th><th>slen</th><th>length</th><th>evalue</th><th>qcovs</th><th>pident</th><th>qstart</th><th>qend</th></tr>
<tr><td>TRINITY_DN33406_ c0_g1_i1</td><td>ref|YP_009407943.1|</td><td>RNA-dependent RNA polymerase, partial [Lasius niger virus 1]</td><td>12,040</td><td>2145</td><td>2150</td><td>0</td><td>53</td><td>86</td><td>7056</td><td>616</td></tr>
<tr><td>TRINITY_DN33406_ c0_g1_i1</td><td>ref|YP_009407942.1|</td><td>Putative capsid protein [Lasius niger virus 1]</td><td>12,040</td><td>487</td><td>488</td><td>0</td><td>12</td><td>82</td><td>9153</td><td>7717</td></tr>
<tr><td>TRINITY_DN33406_ c0_g1_i1</td><td>ref|YP_009407941.1|</td><td>Putative capsid protein [Lasius niger virus 1]</td><td>12,040</td><td>270</td><td>270</td><td>9.65E-171</td><td>7</td><td>91</td><td>9965</td><td>9156</td></tr>
<tr><td>TRINITY_DN33406_ c0_g1_i1</td><td>ref|YP_009407938.1|</td><td>Putative capsid protein [Lasius niger virus 1]</td><td>12,040</td><td>254</td><td>254</td><td>5.71E-158</td><td>6</td><td>92</td><td>11,528</td><td>10,767</td></tr>
<tr><td>TRINITY_DN33406_ c0_g1_i1</td><td>ref|YP_009407939.1|</td><td>Hypothetical protein [Lasius niger virus 1]</td><td>12,040</td><td>269</td><td>270</td><td>7.68E-110</td><td>7</td><td>85</td><td>10,767</td><td>9958</td></tr>
<tr><td>TRINITY_DN33406_ c0_g1_i1</td><td>ref|YP_009407940.1|</td><td>Putative transmembrane protein [Lasius niger virus 1]</td><td>12,040</td><td>85</td><td>86</td><td>6.15E-40</td><td>2</td><td>88</td><td>10,763</td><td>10,506</td></tr>
<tr><td>TRINITY_DN33382_ c1_g1_i1</td><td>ref|YP_009337100.1|</td><td>Hypothetical protein [Hubei picorna-like virus 59]</td><td>10,857</td><td>2904</td><td>1341</td><td>4.22E-72</td><td>35</td><td>26</td><td>10,576</td><td>6737</td></tr>
</table>

that viruses mutate at a high rate and, especially, if pooled insect samples have been used for preparing the RNA-seq library. Therefore, we recommend that the contigs that match viral proteins in the BLASTX search will be aligned using CAP3 [16] in order to merge contigs that differ due to small number (even only one) of mismatches and, additionally, to construct larger contigs based on overlaps that involve a few mismatches. Cap3 produces an output folder and five output files, from which .cap.contigs contains the new merged contigs and cap.singlets contains the contigs with no overlapping sequences with other contigs. Use both .cap.contigs and cap.singlets in further analysis.

```
cap3 virus_contigs_from_blastx.fas
```

7. Carry out a new BLASTX similarity search with the CAP3 contigs and singlets by following the instructions in **step 5**, and use the obtained information, i.e., subject titles of positive hits (stitle) together with their alignment lengths (length) and *e*-values (evalue), for identification of virus-derived sequences and complete genomes. Furthermore, predict open reading frames (ORFs) with NCBI ORF Finder [15] or a similar tool, and perform another BLAST search to find a putative function for each ORF, which will help to confirm and annotate the virus sequences. Note that many ORFs still lack functional annotation, but protein-level match in similarity search is an indication of conserved function.
8. Map the quality-filtered reads to the viral genomes to get insight on the abundance of each virus in your samples. When selecting the mapping software the read length must be taken into account. If the length of the majority of reads is >70 nt you can use, e.g., BWA-MEM [17]; if shorter, then, e.g., Bowtie [18] is suitable. Here, we assume that the reads are >70 nt and give example commands for BWA-MEM.
   (a) First you need to index the virus genomes:

   ```
   bwa index -a is annotated_virus_genomes.fas
   ```

   (b) Then, run the mapping:

   ```
   bwa mem -t 2 annotated_virus_genomes.fas left_
   reads_paired.fq right_reads_paired.fq > output.
   sam
   ```

   (c) Filter the output file according to mapping quality to discard reads that map to multiple locations. Especially viruses that have poly-A tails will have inflated read counts if the mapping quality filtering is skipped:

   ```
   samtools view -q 20 -b output.sam > output_q20.
   bam
   ```

(d) To retrieve the read counts per virus genome, you first need to sort and index the virus genomes:

```
samtools sort output_q20.bam -o output_q20.
sorted.bam
samtools index -b output_q20.sorted.bam
```

(e) Finally, the read counts per virus genome can be obtained:

```
samtools idxstats output_q20.sorted.bam >
stats_per_virus_genome.tsv
```

9. To be able to compare viral read counts between samples, you need to normalize the counts for the number of total filtered reads per sample and the length of the virus genome to get RPKM (reads per kilobase per million) counts. To obtain RPKM counts, you need to divide the number of filtered reads by million to get a scaling factor. Then, divide the read count per virus genome first by the scaling factor (normalization for sequencing depth) and then by the length of the genome in kilobases.

10. It is advisable to set a threshold of viral read fraction out of the total number of filtered reads to be able to separate viruses that truly infect the insect host from those that might just be present in the digested food. For example, we have used a requirement of a minimum of 0.1% of reads mapping to the virus genome, after [3].

11. Active replication of positive-strand viruses in the host can be examined with strand specificity of reads, as positive-strand viruses replicate through negative-strand intermediate. Strand-specificity information of reads derived from negative- and double-stranded viruses gives an indication of transcription efficacy of the genomes. Furthermore, as the double-stranded RNA is targeted by the RNA interference response, strand-specificity information can provide indication of how efficiently the host is able to react to the virus. The strand specificity of the RNA-seq data can be analyzed by running a perl script "examine_strand_specificity.pl" in the Trinity software package. This script examines the distribution of read orientations of the virus sequence. The script also draws an R-dependent violin plot:

```
TRINITY_HOME/util/misc/examine_strand_specificity.
pl output_q20.sorted.bam
```

An example of the script output is shown in Table 2.

3.2 Processing and Analysis of Small RNA (sRNA) Sequence Data

1. The sRNA protocol assumes you have single-end size-selected (read length ≤ 50 nt) sequence data generated by Illumina technology. Again, remove adapter sequences, trim low-quality nucleotides from both ends of the reads, and discard low-quality reads by using, e.g., Trimmomatic [11]:

Table 2
Example of the strandedness analysis results with Trinity script for viruses with negative- (−), positive- (+), or double-stranded (ds) RNA genome

#Transcript	Plus_strand_1stReads	Minus_strand_1stReads	Total_reads	Diff_ratio
Virus1 (+)	61,149	7220	68,369	0.789
Virus2 (+)	53,049	106	53,155	0.996
Virus3 (−)	6448	1238	7686	0.678
Virus4 (−)	1080	2743	3823	−0.435
Virus5 (ds)	1302	398	1700	0.532

```
trimmomatic SE -threads 8 -phred33 SE_reads.fq.gz fil-
tered_SE_reads.fq.gz ILLUMINACLIP:TruSeq3-SE.fa:2:30:10
LEADING:3 TRAILING:3 SLIDINGWINDOW:4:15 MINLEN:15
```

2. It is possible to use VirusDetect pipeline to assemble contigs using the filtered sRNA reads, perform a BLAST search with the contigs against specific virus databases, and map the filtered reads against the virus contigs to get sRNA size distribution for each virus to identify viruses that have been targets of the host RNAi response [19]. However, when we used both RNA-seq and sRNA-seq for the same samples, we were able to assemble complete virus genomes only with the RNA-seq approach since the contigs assembled from sRNA reads were very short (mean length 180 nt). Therefore, we recommend that you use the VirusDetect pipeline if you have only sRNA-seq data. In case you have both RNA-seq and sRNA-seq data, assemble complete virus genomes with the RNA-seq data as above and continue with the next step.
3. Map the sRNA-seq reads against the virus genomes in order to get the sRNA size distribution for each virus. Here, the read length is the most important factor when selecting the mapping software. Now that the read length is <50 nt, e.g., Bowtie [18] is a suitable software:
   (a) First, index the virus genomes (the argument for the <ebwt_base> parameter is a user-selected basename for the index files):

```
bowtie-build -q -f annotated_virus_genomes.fas
<ebwt_base>
```

   (b) Then, map the reads against the virus genomes:

```
bowtie --quiet -v 1 -a --best --strata -S <ebwt_
base> filtered_SE_reads.fq.gz sRNA_output.sam
```

   (c) Filter the output file for mapping quality:

```
samtools view -q 20 -b sRNA_output.sam -o sRNA_
output_q20.bam
```

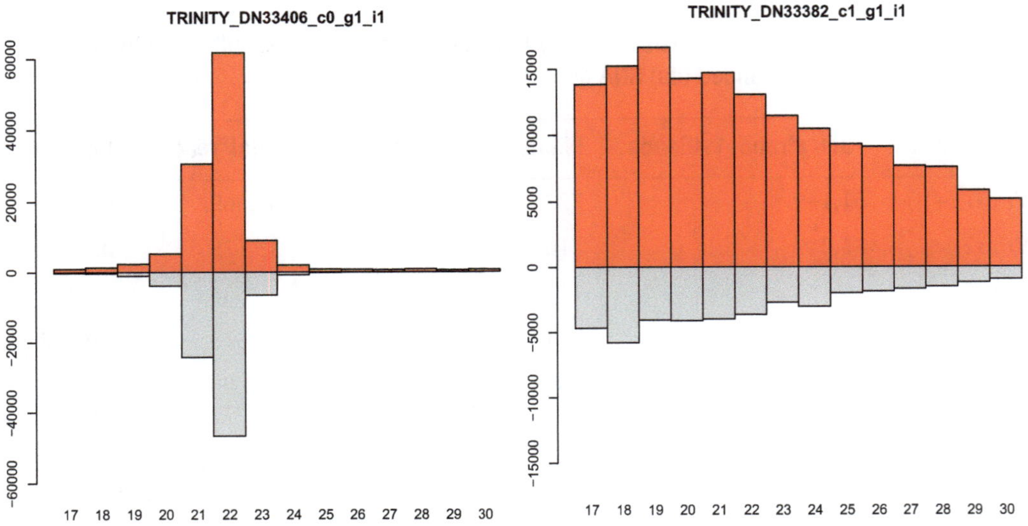

Fig. 1 The sRNA size distribution for two viruses, where the size is on the *x*-axis and the frequency on the *y*-axis. Sense-strand-derived sRNAs are above the *x*-axis and antisense-strand-derived sRNAs below the *x*-axis. (**a**) An example of a distribution from a virus that has been recognized and processed by the host Dicer enzyme to generate mainly 22 nt long small interfering RNAs (siRNAs). The host Dicer recognizes double-stranded RNA, and therefore we expect to see equal representation of sense- and antisense-derived siRNAs. (**b**) An example of virus-derived sRNAs that most probably have been degraded and not processed by the host Dicer enzyme

(d) Sort and index the quality-filtered file:

```
samtools sort sRNA_output_q20.bam -o sRNA_output_
q20_sorted.bam
samtools index -b sRNA_output_q20_sorted.bam
```

4. To find out whether a virus has activated the host RNAi response, you need to visualize the size distribution of sRNAs that map to the virus. An R package ViRome [20] can be used for generating the size distribution and many other charts based on the bam file. In Fig. 1a is a sRNA size distribution of an virus that shows a clear peak at 22 nt, both in the sense and antisense strands, indicating an active RNA interference response against the virus. Figure 1b is an example of sRNA size distribution without a peak at 21–22 nt and uneven representation of sense- and antisense-derived siRNAs suggesting that the virus is able to evade the host RNA interference response.

3.3 Phylogenetic Analysis of the Discovered Viruses

Phylogenetic analysis, together with the information on virus genome organization, can be used to classify viruses and to investigate the relationship of a novel virus with other, similar viruses. The protein sequence of the RNA-dependent RNA polymerase (RdRP) is typically used for the phylogenetic analysis since that is the most conserved protein of RNA viruses, although it should be noted that it gives a restricted picture of the evolutionary history as there is considerable amount of recombination among different viruses and even between viruses and cellular organisms [3]. You may select the viruses for phylogenetic analysis by performing a BLASTP search with the RdRP protein sequence of the novel virus, downloading the protein sequences of selected BLASTP search hits in fasta format and performing the following steps:

1. Multiple sequence alignment of the novel virus RdRP sequence with the selected BLASTX hits using MAFFT [21]

```
mafft --ep 0 --genafpair --maxiterate 1000 input.
fas > alignment.fas
```

2. The alignments tend to contain regions that are difficult to align reliably, so it is recommended that these regions are trimmed with TrimAl [22]

```
trimal -in alignment.fas -out trimmed_alignment.
fas -gt 0.5 -st 0.001 -cons 60 -sgc -scc -sident
```

3. Select amino acid substitution model using ProtTest [23]

```
java -jar prottest-3.4.2.jar -i trimmed_alignment.
fas -o prottest_output.txt -log enabled -all-dis-
tributions -AICC -F
```

4. Reconstruct the phylogeny with PhyML [24] with the command below, where the arguments for the parameters in brackets are obtained in the ProtTest output:

```
phyml -i trimmed_alignment.phy -d aa -m [model]
-f [eq_freq] -v [invariale_sites] -a [gamma]
```

References

1. Berenbaum M (2017) Insect biodiversity—millions and millions. In: Foottit RG, Adler PH (eds) Insect biodiversity: science and society, vol 1, 2nd edn. Wiley-Blackwell, Hoboken, New Jersey
2. Webster CL, Waldron FM, Robertson S et al (2015) The discovery, distribution, and evolution of viruses associated with *Drosophila melanogaster*. PLoS Biol 13(7):e1002210. https://doi.org/10.1371/journal.pbio.1002210
3. Shi M, Lin X, Tian J et al (2016) Redefining the invertebrate RNA virosphere. Nature 540:539–543
4. Viljakainen L, Holmberg I, Abril S et al (2018) Viruses of invasive argentine ants from the European Main supercolony: characterization, interactions and evolution. J Gen Virol 99(8):1129–1140
5. Doublet V, Poeschl Y, Gogol-Döring A et al (2017) Unity in defence: honeybee workers exhibit conserved molecular responses to diverse pathogens. BMC Genomics 18:207. https://doi.org/10.1186/s12864-017-3597-6
6. Gerth M, Hurst GD (2017) Short reads from honey bee (*Apis* sp.) sequencing projects reflect microbial associate diversity. PeerJ 5:e3529. https://doi.org/10.7717/peerj.3529
7. Viljakainen L, Jurvansuu J, Holmberg I et al (2018) Social environment affects the transcriptomic response to bacteria in ant queens. Ecol Evol 8(22):11031–11070
8. Sanjuan R, Nebot MR, Chirico N et al (2010) Viral mutation rates. J Virol 84:9733–9748
9. Nowak MA (1992) What is a quasispecies? Trends Ecol Evol 7:118–121
10. Conesa A, Madrigal P, Tarazona S et al (2016) A survey of best practices for RNA-seq data analysis. Genome Biol 17:13. https://doi.org/10.1186/s13059-016-0881-8
11. Bolger AM, Lohse M, Usadel B (2014) Trimmomatic: a flexible trimmer for Illumina sequence data. Bioinformatics 30:2114–2120
12. Andrews S (2018) FASTQC, A quality control tool for high throughput sequence data.

https://www.bioinformatics.babraham.ac.uk/projects/fastqc/. Accessed 18 April 2018
13. Grabherr MG, Haas BJ, Yassour M et al (2011) Full-length transcriptome assembly from RNA-Seq data without a reference genome. Nat Biotechnol 29:644–652
14. Campillo-Balderas JA, Lazcano A, Becerra A (2015) Viral genome size distribution does not correlate with the antiquity of the host lineages. Front Ecol Evol 3:143. https://doi.org/10.3389/fevo.2015.00143
15. NCBI Resource Coordinators (2017) Database resources of the national center for biotechnology information. Nucleic Acids Res 45:D12–D17
16. Huang X, Madan A (1999) CAP3: a DNA sequence assembly program. Genome Res 9:868–877
17. Li H, Durbin R (2009) Fast and accurate short read alignment with burrows-wheeler transform. Bioinformatics 25:1754–1760
18. Langmead B, Trapnell C, Pop M et al (2009) Ultrafast and memory-efficient alignment of short DNA sequences to the human genome. Genome Biol 10:R25. https://doi.org/10.1186/gb-2009-10-3-r25
19. Zheng Y, Gao S, Padmanabhan C et al (2017) VirusDetect: an automated pipeline for efficient virus discovery using deep sequencing of small RNAs. Virology 500:130–138
20. Watson M, Schnettler E, Kohl A (2013) viRome: an R package for the visualization and analysis of viral small RNA sequence datasets. Bioinformatics 29:1902–1903
21. Katoh K, Standley DM (2013) MAFFT multiple sequence alignment software version 7: improvements in performance and usability. Mol Biol Evol 30:772–780
22. Capella-Gutiérrez S, Silla-Martínez JM, Gabaldón T (2009) trimAl: a tool for automated alignment trimming in large-scale phylogenetic analyses. Bioinformatics 25:1972–1973
23. Darriba D, Taboada GL, Doallo R et al (2011) ProtTest 3: fast selection of best-fit models of protein evolution. Bioinformatics 27:1164–1165
24. Guindon S, Dufayard J, Lefort V et al (2010) New algorithms and methods to estimate maximum-likelihood phylogenies: assessing the performance of PhyML 3.0. Syst Biol 59:307–321

Part VI

Infection Models

Chapter 13

A Workflow for Infection of *Drosophila* with Entomopathogenic Nematodes to Monitor Immune Gene Transcriptional Activity

Christa Heryanto, Eric Kenney, and Ioannis Eleftherianos

Abstract

Drosophila melanogaster has been proven again to be an exceptional experimental tool for studying the efficacy of entomopathogenic nematodes (EPN) as biocontrol agents. Aiming to examine the molecular basis of the insect immune response to EPN infection with or without its mutualistic bacteria, we present a set of established routine techniques for generating axenic culture of infective juveniles of *Heterorhabditis bacteriophora* and *Steinernema carpocapsae* nematodes from their non-axenic culture, followed by a reliable and adaptable infection method of *D. melanogaster* larvae with the EPN stock. Finally, the sensitivity of quantitative real-time PCR method enables us to measure the immune gene transcript levels to explore the dynamics of this complex host-parasite interaction through various timescales and challenge conditions.

Key words *Drosophila*, Entomopathogenic nematodes, *Heterorhabditis*, *Steinernema*, Immunity, Parasitism, Gene expression

1 Introduction

As a field of inquiry, the study of entomopathogenic nematodes (EPN) has gained notable momentum in recent years due largely to their known efficacy as biocontrol agents against insect pests as well as the beckoning potential of these parasites to be made even more effective through biological manipulation [1]. Naturally, any approach to understanding this system is consistently challenged by the complexity of the host-parasite interactions that determine whether it will be the assorted virulence factors of the nematode or the diverse immune mechanisms of its insect host that finally emerge victorious in their efforts to establish control of the hemocoel environment. The identification of significant patterns of immunity or virulence in a system displaying this degree of complexity can be a daunting task, but a number of studies have shown distinct successes through the use of qPCR

Federica Sandrelli and Gianluca Tettamanti (eds.), *Immunity in Insects*, Springer Protocols Handbooks,
https://doi.org/10.1007/978-1-0716-0259-1_13,

and RNA-seq methodologies that allow the user to accurately and efficiently survey the transcriptional responses of their organism of interest.

The utility of a transcriptomic study is bolstered substantially by targeting an organism with higher degree of available genetic information, and so many studies have relied on *Drosophila* as a model host in which to examine the molecular basis of the immune response of an insect to an EPN infection. The signaling pathways of *Drosophila* have been described with at least enough resolution that the primary immune signaling pathways including Toll, Imd, JAK/STAT, JNK, and TGF-β each have known genes that reliably report their activity and can be used to indicate their level of induction in response to an EPN challenge [2–4]. Measuring pathway activity can be achieved with a straightforward, established workflow involving infection, RNA extraction, reverse transcription, and qPCR, though it should be noted that the qPCR step should involve careful optimization of primer design, annealing temperatures, and template concentration to achieve accurate and repeatable results [5]. If a research question is broader in scope and the individual measurement of genes is not feasible, the qPCR step of this workflow can easily be supplanted with RNA-seq analysis [6]. Conveniently as well, the results of these assays can garner information with regard to both sides of the host-parasite interaction by highlighting immune pathways activated in response to EPN infection as well as pathways suppressed by nematode parasites to further undermine insect immunity. Moreover, inferences from an EPN-*Drosophila* interaction can potentially be expanded to other host-parasite systems as some classes of virulence factors may be shared across diverse groups of nematodes [7].

Here, the methods of this workflow have been detailed with a focus on established techniques for generating stocks of nematodes both with and without their mutualistic bacteria as well as a reliable and adaptable method for infecting *Drosophila* larvae with said EPN stocks. When coupled with the ease and sensitivity of measuring gene expression via qPCR, this system can be of ineffable value in unraveling the complexity of host-parasite interactions through a wide variety of timescales and infection conditions.

2 Materials

2.1 Generating Symbiotic Nematode Infective Juveniles (IJs)

1. *Heterorhabditis bacteriophora* or *Steinernema carpocapsae* IJs (up to 6 weeks old).
2. Waxworm (*Galleria mellonella*).
3. 10-cm and 15-cm petri dish.
4. 15-cm and 15-cm filter paper.

2.2 Generating Axenic Nematode IJs

2.2.1 Heterorhabditis bacteriophora IJs

1. *Photorhabdus temperata* Ret16 bacterial stock.
2. 22G needle and 1-ml syringe.
3. 15–20 waxworms (*G. mellonella*).
4. Symbiotic or axenic *H. bacteriophora* IJs.

2.2.2 Steinernema carpocapsae IJs

1. Nutrient Broth (Difco).
2. Yeast.
3. Agar.
4. 0.98 M $MgCl_2$, autoclaved separately.
5. 7.3% corn syrup (v/v), autoclaved separately.
6. Corn oil, autoclaved separately.
7. 30 mg/ml kanamycin, 0.2-μm filter-sterilized.
8. 50 mg/ml ampicillin, 0.2-μm filter-sterilized.
9. Partitioned petri dish/split plate/I-plate (e.g., VWR 25384-310).
10. LB broth, supplemented with 30 μg/ml kanamycin and 50 μg/ml ampicillin.
11. *Xenorhabdus nematophila ΔrpoS* bacterial stock.
12. Symbiotic *Steinernema carpocapsae* IJs.

2.3 Infection Method of Drosophila Larvae by Co-incubation with Nematode IJs

1. 1.5% agarose gel.
2. 96-well microtiter plate.
3. Sterile distilled water.
4. PCR plate film, cut for each row.
5. Needle to puncture PCR plate film.
6. Fly vials with third instar larvae.
7. Surface-sterilized nematode IJs (e.g., 100 IJs/10 μl).

2.4 Gene Transcription Analysis

1. RNA isolating reagent (e.g., TRIzol®) and other reagents as specified in the manufacturer's protocol.
2. 5 mg/ml glycogen.
3. Motorized homogenizer with pestle.
4. Larval samples in 1.5-ml vials (about four larvae per vial).
5. cDNA synthesis kit of choice.
6. Primers targeted at gene of interest and housekeeping gene.
7. qRT-PCR kit of choice.

3 Methods

3.1 Generating Symbiotic Nematode Infective Juveniles (IJs)

1. Line a 10-cm petri dish with a sheet of filter paper and pick about 10–15 waxworms (Fig. 1a).
2. Dispense on the waxworms using 1-ml pipette 2 ml solution of approximately 25–50 IJs per 10 μl of water. Leave the petri dish in the dark (*see* **Note 1**).
3. Every 2 days, add 1–2 ml of water depending on the moisture of the filter paper. The waxworms will typically succumb to the infection within 48 h.
4. Ten days after the IJs are applied to the waxworms, prepare White's water traps [8] (Fig. 1b, c). Transfer the dead waxworms onto the unsubmerged portion of the water trap filter paper carefully. Add tap water onto the bottom dish and use the cap of a 15-ml tube as a spacer for ventilation (*see* **Note 2**).
5. Once the water around the small petri dish is cloudy with nematodes, transfer the new IJs into a cell culture flask, such as T25 or T75 flasks using a pipette.
6. Add pure water until the desired density is reached, up to about 40% of the volume (Fig. 1d). Avoid overcrowding. Store horizontally.
7. Replace the removed water and repeat **steps 5** and **6** until IJs cease to emerge from the infected waxworms.

3.2 Generating Axenic Nematode IJs

3.2.1 Heterorhabditis bacteriophora IJs

1. Culture single bacterial colonies on a MacConkey plate, incubated for 2–3 days at 28 °C.
2. Inoculate a colony in LB broth at 28 °C and incubate overnight at 28 °C in a shaking incubator.
3. Wash 1 ml of overnight culture with 1× PBS. Dilute 10× in 1× PBS. Leave on ice.
4. Dip the waxworms into 70% ethanol, dry on paper towel, and collect in 50-ml tube.
5. Submerge the tube in ice for 20 min to anesthetize.
6. Line the top and bottom halves of a 10-cm petri dish with filter paper.
   (a) Use the lid lined with filter paper as a base for injection.
   (b) Dampen the filter of the bottom dish and place on ice to allow for recovery of the injected waxworms.
7. Place a droplet (50 μl) of ice-cold bacteria culture on the parafilm and fill in the syringe. Push the plunger slightly to evacuate air at the tip of the needle.
8. Under the stereoscope, hold down a waxworm close to the posterior end of the larva (Fig. 2).

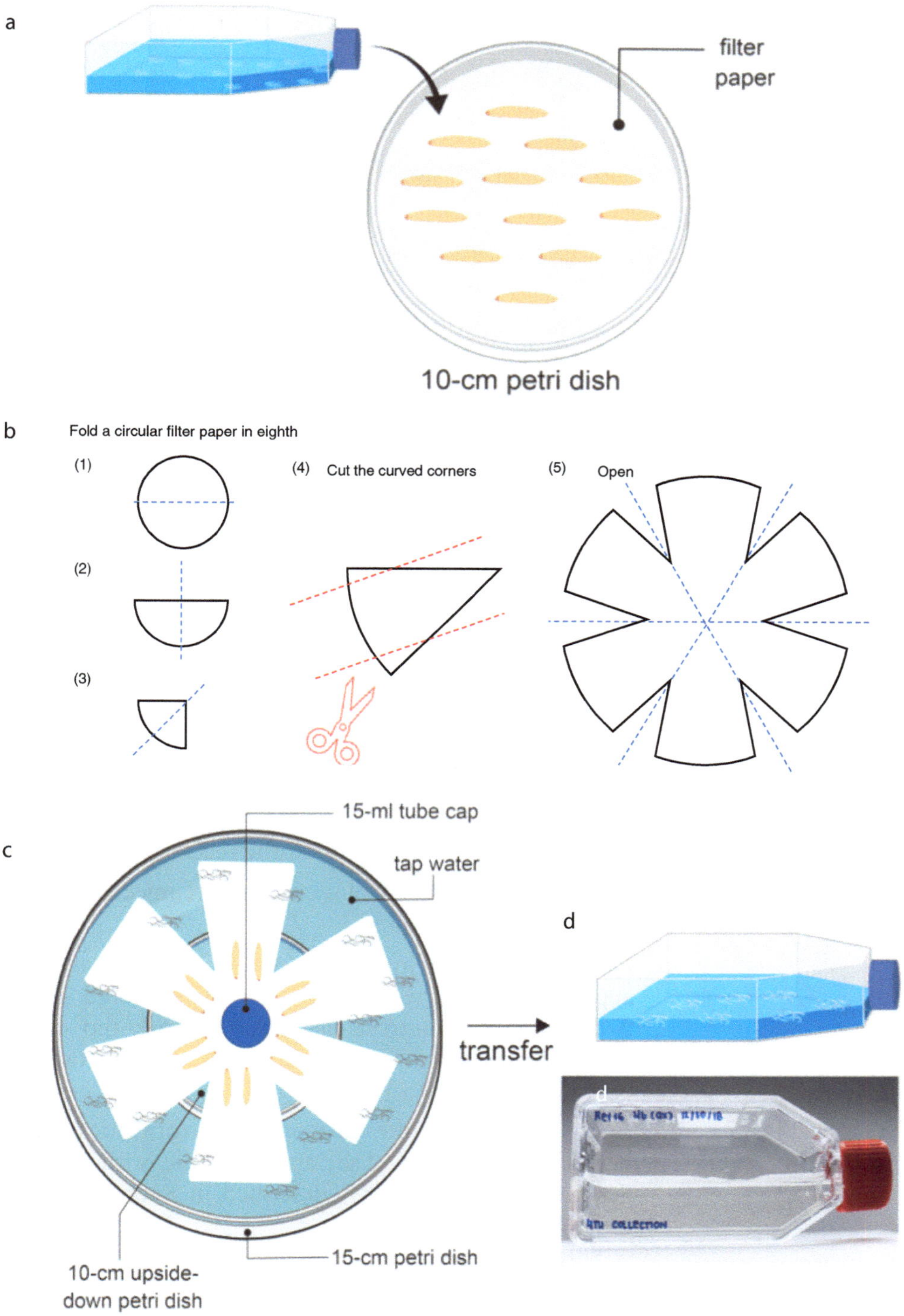

Fig. 1 Generating symbiotic nematode infective juveniles. (**a**) Infection of waxworms with symbiotic infective juveniles. (**b**) Filter paper folding for water trap. (**c**) Top view of the water trap with dead waxworm carrying infective juveniles 12 days after infection. The new generation of nematode infective juveniles will exit the host and enter the water, which is then collected into the new T75 flask for storage. (**d**) Collected infective juveniles in a small flask. Images are made using Biorender graphic software (https://biorender.com)

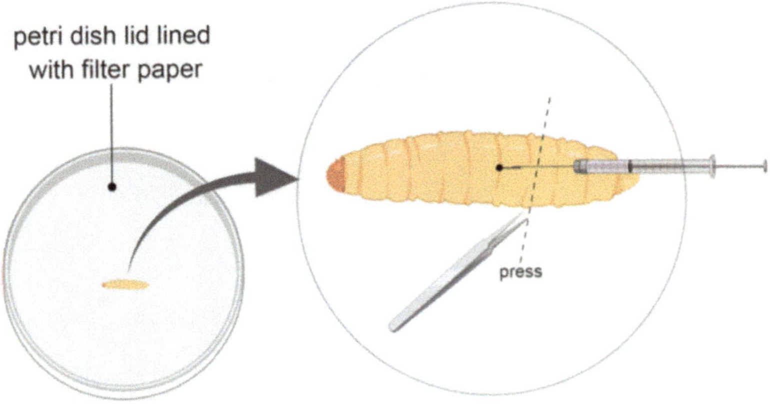

Fig. 2 Injecting a waxworm with *Photorhabdus temperata* Ret16 bacteria. Pressing the posterior area engorges its body to present a turgid injection surface and prevents sliding during injection. The injection angle should be as shallow as possible to avoid damaging internal structures. Images are made using Biorender graphic software (https://biorender.com)

9. Aim to inject into the dorsal side of the thorax behind the legs, preferably at the junction between two segments, right underneath the cuticle as parallel to the waxworm as possible to minimize internal damage. It is natural to sometimes have a droplet of hemolymph bled out when the waxworm is pierced, but the entire 50 μl of the liquid should penetrate under the cuticle.
10. Place the injected waxworm on the recovery dish.
11. Repeat **steps 7–9** until all waxworms are injected. Leave waxworms on ice for 5 min.
12. Place the petri dish in a drawer and wet with tap water if filter appears almost dry. Galleria caterpillars should die within 2 days and turn into a characteristic brick red color after 3–4 days. If the waxworms are brown, the infection has not proceeded properly.
13. After 7 days, move the waxworms that turn brick red onto a new filter-lined petri dish, and continue with "Generating Symbiotic Nematode IJs" described in Subheading 2.1 (*see* **Note 3**).

    Surface-sterilizing *H. bacteriophora* IJs:
14. Pellet 100 μl each of symbiotic *H. bacteriophora* and candidate-axenic *H. bacteriophora* in a 1.5-ml vial.
15. Add 5% bleach to each vial and invert. Incubate for 10 min. Spin at 17,900 × *g*.
16. Working in biosafety cabinet, wash five times with 1 ml sterile water.

17. Decant.

    Verifying axenicity:

18. Add 400 μl of sterile water to symbiotic *H. bacteriophora* and candidate-axenic *H. bacteriophora*.
19. Add the nematodes on top of the waxworms in petri dishes and label them accordingly.
20. Incubate in the dark and check periodically for appearance of red color. Waxworms infected with symbiotic *H. bacteriophora* will turn red after 2 days. This is due to the presence of mutualistic *Photorhabdus* bacteria secreting chemicals during infection that turn the waxworm red. Candidate-axenic *H. bacteriophora* is confirmed as axenic when the waxworms do not turn red by day 4 post infection since *P. temperata* Ret16 is absent in the gut of the nematodes.

3.2.2 *Steinernema carpocapsae* IJs

1. Preparing lipid agar (300 ml makes about 20 split plates):
   - (a) Mix 2.4 g of Nutrient Broth, 4.5 g of yeast, and 1.5 g of agar in 267 ml deionized water. Autoclave.
   - (b) Supplement with 3 ml of 0.98 M $MgCl_2$, 28.8 ml of 7.3% corn syrup, and 1.2 ml of corn oil.
   - (c) Add 300 μl of 30 mg/ml kanamycin and 300 μl of 50 mg/ml ampicillin (0.2-μm filter-sterilized).
   - (d) Pour on one side of the partitioned petri dish/split plate.
2. Preparing *X. nematophila ΔrpoS* bacterial lawn:
   - (a) Grow *X. nematophila ΔrpoS* overnight in 2 ml of LB/kan/amp in a shaking incubator at 28 °C.
   - (b) Use 250 μl of the overnight culture to inoculate 5 ml of LB/kan/amp in a shaking incubator.
   - (c) Pipette 100 μl of the culture on lipid agar plates and spread over the entire plate with a sterile inoculating loop.
   - (d) Incubate the plates for 24 h at 28 °C.
3. Surface-sterilizing *S. carpocapsae* IJs:
   - (a) Pellet by centrifugation 1 ml of concentrated IJs (by leaving the flask at an angle for 10 min) at 17,900 × *g* for 10 s. Obtain about 0.5 ml of IJ pellet.
   - (b) Decant water and add 1 ml of 1% bleach. Invert tubes to mix thoroughly; let stand for 1 min (*see* **Note 4**). Spin again for 10 s. Decant bleach solution.
   - (c) Wash the nematodes with 1 ml of sterile distilled water five times to remove the bleach residue.
   - (d) Obtain IJ counts.

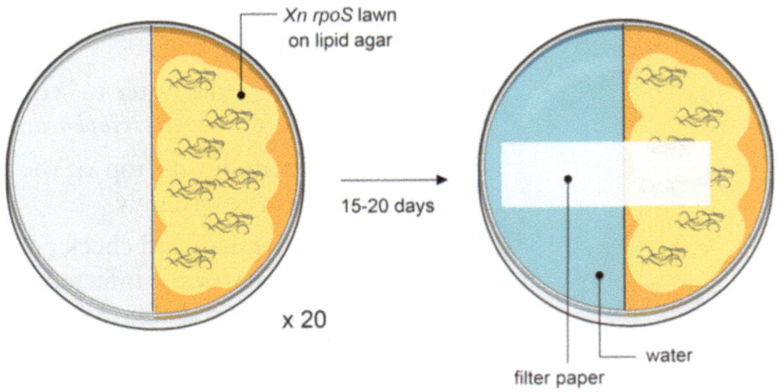

Fig. 3 Generating axenic *Steinernema carpocapsae* using *Xenorhabdus nematophila rpoS* mutant lawn. A split plate with *X. nematophila rpoS* on lipid agar facilitates the development of infective juveniles into adulthood, which will lay eggs and hatch into the new generation. Once the plate is mostly populated with newly generated infective juveniles, water and filter paper are placed as pictured to collect the worms. Images are made using Biorender graphic software (https://biorender.com)

4. Developing *S. carpocapsae* IJs:
   (a) Pipette about 1000 IJs onto each lipid agar split plate in random droplets (Fig. 3, left picture). Store the plates at room temperature (25 °C) in a dark, humidified cabinet. Line the cabinet with damp paper towels (*see* **Note 5**).
   (b) Once IJs are present without any other larval stages after approximately 15–20 days, fill the other side of the plate with water, and lay a strip of filter paper across the middle (Fig. 3, right picture). Collect the IJ suspension from the split plates into T75 flasks (*see* **Notes 3** and **6**).
5. Verifying axenicity of *S. carpocapsae* IJs by Colony-Forming Unit (CFU):
   (a) Surface-sterilize first round, second round, and symbiotic *S. carpocapsae* IJs as described in **step 3**.
   (b) Homogenize these IJs with sterile pestles. Spin down the homogenate, decant, and spread onto LB agar. Incubate the plates at 28 °C.
   (c) Count the CFU for each plate. Axenic samples should not form any colonies.
6. Verifying axenicity of *S. carpocapsae* IJs by PCR:
   (a) Surface-sterilize first round, second round, and symbiotic *S. carpocapsae* IJs as described in **step 3**.
   (b) Extract DNA from the homogenate.

(c) Conduct PCR diagnostics using the primers listed below (annealing temperature, 61 °C):

XptA F: 5′-GCCTGGAAAGAGTGGACGAA-3′.

XptA R: 5′-GTAAGACCAAGGGGCACTCC-3′.

(d) Visualize the amplified fragments on 1.5% agarose gel. Axenic samples should not form any bands (amplicon found at 231 bp).

3.3 Infection Method of Drosophila Larvae by Co-incubation with Nematode IJs

Prior to sampling infected *D. melanogaster* larvae for transcriptional analysis [9], a survival experiment may be conducted to determine collection time points that suit the needs of the experiment. Three time points should be chosen based on the percent survival of *D. melanogaster* in response to a certain treatment:

- Early time point: 100% survival, usually between 0 and 6 h post infection.
- Mid time point: at a time point in which approx. 50% survival is observed.
- Late time point: at a time point in which approx. 1–10% survival is observed or when survival becomes constant and no more deaths are observed.

For non-infected controls, *D. melanogaster* larvae are incubated with water of the same source as that in which the nematodes were suspended. If a *D. melanogaster* mutant line is used, the correct background line should be included as a control in the experiment.

Experimental variation can be addressed by performing three independent experiments with two biological replicates per experiment as a standard practice. For sampling for qRT-PCR analysis, two biological replicates are established by pooling at least four larvae of the same treatment in two separate tubes. This setup is repeated three times with a different batch of larvae to produce independent experiments.

1. Fill each well of the 96-well plate with 100 μl of 1.5% agarose gel. Cool at 4 °C for approximately 20 min.
2. Using a small paint brush, collect third instar *D. melanogaster* larvae onto a filter paper on a petri dish. Rinse the food debris off the larvae body with a small drop of sterile water (Fig. 4).
3. Place larvae into the wells of the 96-well plate.
4. Add 10 μl of IJ suspension per well or 10 μl of sterile distilled water as control.
5. Seal the wells of the 96-well plate with PCR film one row at a time, and puncture each well with a needle for ventilation (*see* **Note 7**).

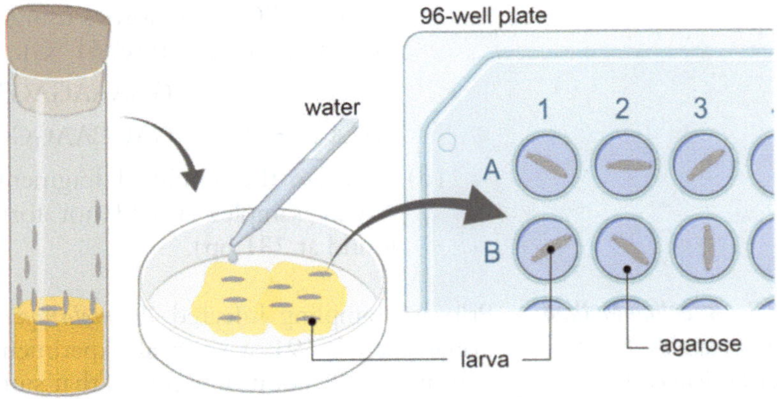

Fig. 4 Transfer of *Drosophila melanogaster* third instar larvae into 96-well plate for infection with nematode infective juveniles. Larvae are collected with paint brush from a fly vial onto a petri dish lined with a filter paper and rinsed off with a drop of water. Once the food is rinsed off, these larvae are placed one by one into each well of the 96-well plate. Images are made using Biorender graphic software (https://biorender.com)

6. For survival experiments, count the number of live and dead larvae by scoring them based on their motion.
7. For gene expression studies (*see* Subheading 3.4), collect at least four larvae per treatment using a paint brush in 1.5-ml microcentrifuge tube and store at −80 °C.

3.4 Gene Transcriptional Analysis

3.4.1 Tissue Lysis and Total RNA Isolation

Liquid–liquid RNA isolation based on acid guanidine isothiocyanate-phenol-chloroform method [10], e.g., TRIzol®, is preferred based on its ease of use and high yield as compared to commercial extraction kits. To completely lyse stored samples, frozen larval samples are homogenized in TRIzol® solution using a motorized tissue homogenizer with autoclavable pestle attachment (Fig. 5, tissue lysis) until there are no identifiable *D. melanogaster* structures. 500 μl of TRIzol® is typically sufficient for five adult flies or larvae. BCP is also used as a safer alternative to chloroform. 1 μl of 5 mg/ml glycogen is also added into the new tube prior to transferring the clear upper phase. The rest of the procedures are carried out as stated in the manufacturer's protocol. RNA dissolved in solution should be kept on ice whenever possible to prevent degradation of the sample (*see* **Note 8**).

3.4.2 cDNA Synthesis

Prior to performing the reverse transcription reaction, we normalize the amount of RNA throughout the sample set (e.g., 1000 ng per reaction). If RNase inhibitor is not included in the kit or is not part of the mix, RNase inhibitor is added to the reaction as per the manufacturer's instruction. The resulting cDNA is then diluted

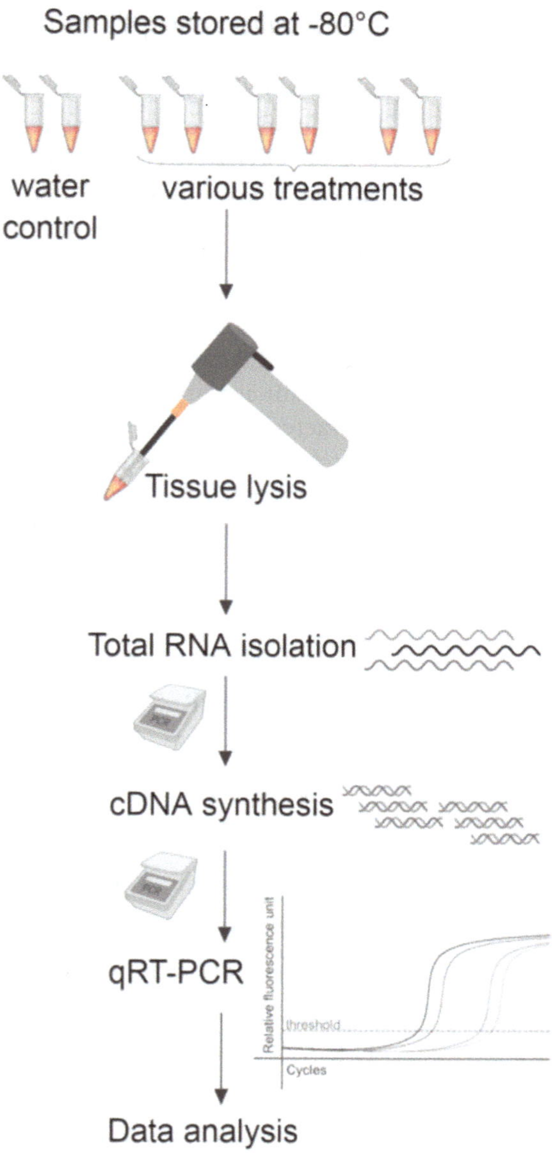

Fig. 5 Workflow of gene expression analysis using quantitative real-time PCR method. Images are made using Biorender graphic software (https://biorender.com)

10× for quantitative real-time PCR to reduce any reaction interference from the components of the cDNA reaction mixture.

3.4.3 Quantitative Real-Time PCR

While most of the general rules for primer design still apply, the size of the target amplicon is optimal between 50 and 150 bp. Each primer pair of a gene of interest amplifying one sample must be accompanied by a separate reaction with the reference gene's primer pair in one qRT-PCR run, both in two technical replicates (two separate wells) as depicted in Fig. 6a, under Sample. *D. melanogaster Ribosomal protein L32* gene is a standard reference gene [11] in studying transcript level of other genes of interest (*rpL32* F: 5′-GATGACCATCCGCCCAGCA-3′; *rpL32* R: 5′-CGGACCG ACAGCTGCTTGGC-3′). Primers for a number of genes tied fundamentally to the immune response are listed below according to their associated pathways. Imd pathway activity is commonly monitored through the expression of two antimicrobial peptides, *Diptericin* and *Cecropin*, which are active against Gram-negative bacteria. Likewise, the activity of the Toll pathway, with its predominant activity against Gram-positive bacteria and fungi, can be characterized with expression profiles for *Drosomycin* and *Metchnikowin*. For the general stress-responsive JAK-STAT pathway, primers may target the genes for *TotA* and *TotM*, which encode secreted proteins expressed in response to JAK-STAT activation.

Imd:

Diptericin F: 5′-GCTGCGCAATCGCTTCTACT-3′.

Diptericin R: 5′-TGGTGGAGTTGGGCTTCATG-3′.

Cecropin F: 5′-TCTTCGTTTTCGTCGCTCTC-3′.

Cecropin R: 5′-CTTGTTGAGCGATTCCCAGT-3′.

Toll:

Drosomycin F: 5′-GACTTGTTCGCCCTCTTCG-3′.

Drosomycin R: 5′-CTTGCACACACGACGACAG-3′.

Metchnikowin F: 5′-TCTTGGAGCGATTTTTCTGG-3′.

Metchnikowin R: 5′-AATAAATTGGACCCGGTCTTG-3′.

JAK-STAT:

TotA F: 5′-GAAGATCGTGAGGCTGACAAC-3′.

TotA R: 5′-GTCCTGGGCGTTTTTGATAA-3′.

TotM F: 5′-GCTGGGAAAGGTAAATGCTG-3′.

TotM R: 5′-AGGCGCTGTTTTTCTGTGAC-3′.

Once the fold changes are calculated, the result can be presented as illustrated in Fig. 6b, where different time points (early, mid, late) are usually expressed in hours or days post infection. When *D. melanogaster* mutant lines are used, background strain is included to contrast the effect. Finally, using statistical analysis software, such as GraphPad Prism, the significant differences in transcriptional activity in response to EPN challenges at each time point can be automated.

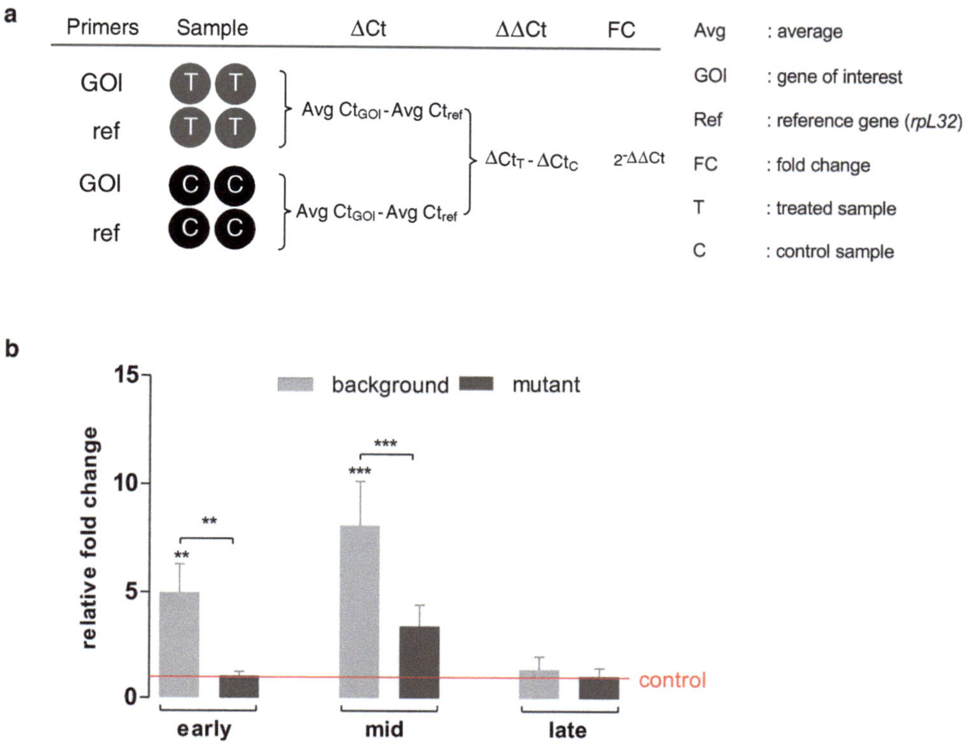

Fig. 6 (**a**) Using $\Delta\Delta C_t$ method [12] to analyze the transcript-level change of a gene of interest relative to the reference gene with two technical replicates. (**b**) Example of transcriptional analysis data presentation in a treated *Drosophila melanogaster* mutant line at early, mid, and late time points using GraphPad Prism. Red line indicates the gene transcript level for the uninfected control which is set at 1. Background and mutant gene transcript levels are normalized to the control value

4 Notes

1. Concentrate the existing IJs by leaving the T75 flask upright for 10 min, allowing the IJs to collect at the bottom of the flask.
2. At this point, the waxworm body will be soft and fragile. Transfer them carefully by loosening them slowly from the filter paper with forceps. Tap water is used as the demineralized or deionized water causes the nematodes to clump. Maintain the water level when necessary as it will evaporate over time depending on room humidity. Level of water should reach at least half the height of the petri dish.
3. If using symbiotic IJs, repeat the process again from **step 1** using the first round of newly generated IJs.
4. Prolonged bleach incubation will lead to IJ death.
5. Check the paper towels every other day to make sure they stay wet. Add water if necessary. A water bath may also be used at the bottom of the cabinet as humidifier.

6. Check under a stereoscope every few days to track the development of the nematodes.
7. USA Scientific PCR film (Cat. No.2921-0000) is preferred for its low stickiness to larvae, while still preventing them to crawl out.
8. Use gloves at all times to prevent RNase contamination and as safety precaution.

References

1. Lacey LA, Grzywacz D, Shapiro-Ilan DI, Frutos R, Brownbridge M, Goettel MS (2015) Insect pathogens as biological control agents: Back to the future. J Invertebr Pathol 132:1–41
2. Patrnogic J, Heryanto C, Eleftherianos I (2018) Transcriptional up-regulation of the TGF-beta intracellular signaling transducer mad of *Drosophila* larvae in response to parasitic nematode infection. Innate Immun 24:349–356
3. Yadav S, Gupta S, Eleftherianos I (2018) Differential regulation of immune signaling and survival response in *Drosophila melanogaster* larvae upon *Steinernema carpocapsae* nematode infection. Insects 9:E17
4. Hallem EA, Rengarajan M, Ciche TA, Sternberg PW (2007) Nematodes, bacteria, and flies: a tripartite model for nematode parasitism. Curr Biol 17:898–904
5. Taylor SC, Nadeau K, Abbasi M, Lachance C, Nguyen M, Fenrich J (2019) The ultimate qPCR experiment: producing publication quality, reproducible data the first time. Trends Biotechnol 37(7):761–774
6. Yang Y, Wen Y, Cai YN et al (2015) Serine proteases of parasitic helminths. Korean J Parasitol 53:1–11
7. Yadav S, Daugherty S, Shetty AC, Eleftherianos I (2017) RNAseq analysis of the *Drosophila* response to the entomopathogenic nematode *Steinernema*. G3 (Bethesda) 7:1955–1967
8. White GF (1927) A method for obtaining infective nematode larvae from cultures. Science 66:302–303
9. Yadav S, Shokal U, Forst F, Eleftherianos I (2015) An improved method for generating axenic entomopathogenic nematodes. BMC Res Notes 8:461
10. Chomczynski P, Sacchi N (1982) Single-step method of RNA isolation by acid guanidium thiocyanate-phenol-chloroform extraction. Anal Biochem 162:156–159
11. Matta BP, Bitner-Mathé BC (2011) Getting real with real-time qPCR: a case study of reference gene selection for morphological variation in *Drosophila melanogaster* wings. Dev Genes Evol 221:49–57
12. KJ L, Schmittgen TD (2001) Analysis of relative gene expression data using real-time quantitative PCR and the 2(−Delta Delta C(T)) method. Methods 25:402–408

Chapter 14

Oral Infection in a Germ-Free *Bombyx mori* Model

Daniel Brady, Alessio Saviane, Ottavia Romoli, Gianluca Tettamanti, Federica Sandrelli, and Silvia Cappellozza

Abstract

Pathogens of the silkworm *Bombyx mori* reduce silk crop quality and quantity, causing significant economic losses to silkworm rearers and the silk industry globally. In order to combat microbial diseases at the agricultural level, it is informative to characterize the host immune responses activated during infection in environmentally controlled conditions. While conventional silkworm rearing is dependent on the seasonality of mulberry trees, in the field of scientific research, recent developments such as artificial diets have resulted in consistent and controlled rearing conditions throughout the year. In this chapter, we describe protocols to perform oral infection experiments in a simplified germ-free silkworm model, reared on artificial diet. Also, we provide simple assays to monitor the activation of the immune response after oral infection, including the evaluation of the pathogen passage from the gut into the hemolymph, the change in the number of hemocytes, the actual rate of melanization, and the antimicrobial activity kinetics of the hemolymph during infection. These standardized protocols will enable the reporting of comparable datasets for *B. mori* host-pathogen interaction among research groups.

Key words *Bombyx mori*, Germ-free rearing condition, Oral infection, Host-pathogen interaction, Midgut, Innate immunity

1 Introduction

The silkworm *Bombyx mori* is an important organism for its economic value and as a model organism for Lepidopteran genetics. *B. mori* has been completely domesticated over the past 5000 years, and critical environmental conditions that contribute to its successful development, such as temperature, relative humidity, and photoperiod, have been well-characterized. However, the control of silkworm diseases remains one of the greatest goals in silk production; in China, the world's largest silk exporter, epidemic outbreaks result in a loss of about 10% of the silk cocoon yield every year [1]. Therefore, characterizing silkworm-pathogen interactions may help to develop practices that will reduce these losses.

Daniel Brady and Alessio Saviane equally contributed to this work.

Federica Sandrelli and Gianluca Tettamanti (eds.), *Immunity in Insects*, Springer Protocols Handbooks,
https://doi.org/10.1007/978-1-0716-0259-1_14,

Like other insects, *B. mori* relies on innate immune responses to counteract pathogenic infections. The primary defenses in *B. mori* are cuticle and peritrophic matrix (PM), physical barriers exposed to microorganisms within the environment, and the alimentary canal, respectively [2, 3]. In the silkworm, the most common route of infection follows ingestion of contaminated diets, particularly in crowded rearing conditions [4, 5]. As a consequence of oral infection, a pathogen can pass from the alimentary canal into the insect's body cavity, the hemocoel, systemically invading the host. The hemocoel is filled with hemolymph, the insect "blood" containing circulating hemocytes. In the hemolymph a rapid and dynamic immune response occurs, characterized by both cellular and humoral components.

The cellular response is mainly triggered by hemocytes, which activate defense processes such as phagocytosis, encapsulation, and nodulation. The humoral response comprises the synthesis of melanin, which in turn contributes to the pathogen encapsulation, and the production of several immune effectors. These include lysozyme and antimicrobial peptides (AMPs), which are synthesized mostly in the fat body and released into the hemolymph after proteolytic maturation [5, 6]. Although to a lesser extent, AMPs are also produced by hemocytes and at cuticle and gut levels, during local infections.

B. mori strains show variable susceptibility to different pathogens after oral infections. This is likely due to variations among strain capabilities to activate the multiple components of the innate immune response [7, 8]. The outcome of silkworm infections can also be influenced by the presence of different microbes in the rearing environment and by the composition of the larval microbiota [9, 10]. Additionally, the quality of mulberry leaves directly affects the physiology of *B. mori* [11, 12]. These latter aspects, when not well-characterized, act as confounding factors in the study of silkworm pathologies and limit the reproducibility of silkworm infection experiments among laboratories. To counteract these constraints, we describe protocols to perform oral infection experiments in a simplified germ-free silkworm model reared on artificial diet. We also provide several simple assays to monitor the infection and the activation of the immune response after oral exposure, including the evaluation of the pathogen passage from the gut into the hemocoel, the change in number of hemocytes, the actual rate of melanization, and the antimicrobial activity kinetics of the hemolymph during the progression of the infection. Finally, we describe an appropriate method to prepare gut samples for the morphological and ultrastructural analyses of the PM.

These protocols can be performed with any *B. mori* strain infected with a pathogen of interest. *Enterococcus mundtii*, a Gram-positive bacterium causing flacherie disease in silkworms reared on artificial diet [7, 13] is used as the example microorganism throughout these procedures.

2 Materials

2.1 Silkworm Strains

Pure and hybrid silkworm strains are usually employed in infection experiments according to local availability. In general, silkworm hybrids are more resistant/tolerant to infections, while pure lines are more sensitive. Very often polyvoltine/bivoltine strains are used in laboratories, as they lay non-diapausing eggs and they can be repeatedly employed in subsequent experiments all year-round. However, in the case of research on diseases, the geographical origin and the inbreeding level influence the sensitivity of a particular silkworm strain to a specific microorganism. Therefore, to have a general character, immunity studies should be carried out on different strains.

2.2 Equipment and Reagents to Obtain Germ-Free Silkworm Cultures on Artificial Diet

1. Autoclaved containers (Fig. 1a).
2. Fine head paintbrush and forceps (sterilized in 70% ethanol) to transfer larvae.
3. Sterile tissue paper (28 × 28 cm, folded twice).
4. Laminar flow hood.
5. 70% ethanol.
6. 95% ethanol.
7. 0.5% sodium hypochlorite (in water).
8. Preparation of artificial diet:
   (a) Prepare 25% mulberry diet powder for hybrid silkworms or 40% mulberry diet powder for pure strains; *see* Table 1 for component quantities (*see* **Note 1**).
   (b) Thoroughly mix artificial diet powder with distilled water; quantities and hydration rates vary with larval instars and are indicated in Table 2 (*see* **Note 1**).

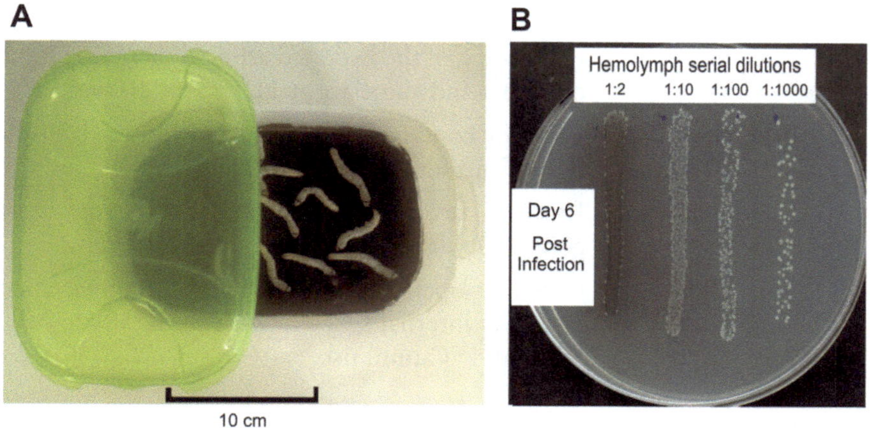

Fig. 1 *E. mundtii* oral infection in silkworms. (**a**) Autoclavable sealable container used to prepare artificial diet (size, 28 cm × 16 cm × 8 cm); scale bar indicates diet container length and width; (**b**) PCA plate showing *E. mundtii* colonies grown on serially diluted hemolymph extracted from an *E. mundtii*-infected *B. mori* larva, 6 days postinfection

Table 1
Components and quantities of dry artificial mulberry diet powder

Ingredients	25% (g/100 g)	40% (g/100 g)
Dried mulberry leaf powder	25	40
Defatted soybean meal	36	29
Wheat meal	15	8.9
Cornstarch	5.4	3.9
Soybean fiber	5	5
Citric acid	4	4
Ascorbic acid	2	2
Salt mixture	3	3
Agar	4.2	4.2
Vitamin B	0.4	0.4

Table 2
Quantities of dried diet powder and rehydration ratio required for different larval stages

Instars	Artificial diet (g)	Powder rehydration rate (g:mL)	Silkworm number	Container surface (cm^2)	Container height (cm)
I–III	40	1:2.2	160	400	5
IV	70	1:2.0	50–60		8
V	60	1:2.0	10–15		8

(c) Pour 180 g of the hydrated artificial diet into an autoclavable container and autoclave at 117 °C for 40 min (*see* **Note 1**).

(d) Cool the container to room temperature (RT) and transfer to 4 °C upside down for up to 7 days.

(e) In a laminar flow, sterilize the external surface of the diet container with 70% ethanol, before opening.

(f) Carefully open the container, and remove condensation from the internal surfaces with sterile tissue (*see* **Note 2**). Store at 4 °C until use.

2.3 Oral Infections of Germ-Free Silkworms with Bacteria

1. Laminar flow hood.
2. UV/VIS spectrophotometer.
3. Bacteriological Petri dishes.
4. 15-mL sterile disposable tubes.

5. 1.5-mL sterile microcentrifuge tubes.
6. 1.5-mL disposable cuvettes.
7. *E. mundtii* stock solution (25% glycerol, maintained at −80 °C).
8. Sterile phosphate-buffered saline (PBS): NaCl (137 mM), KCl (2.7 mM), Na_2HPO_4 (10 mM), KH_2PO_4 (1.8 mM), pH 7.4.
9. Culture media: Plate Count Agar (PCA). Dissolve peptone (0.5%, 2.5 g), yeast extract (0.25%, 1.25 g), glucose (0.1%, 0.5 g), agar (1.5%, 6 g) in 350 mL of pure H_2O, mix, and bring to a final volume of 500 mL. Transfer to a 1-L Schott bottle and autoclave. Cool the medium to 55 °C and pour into Petri dishes inside the laminar flow hood. PCA broth is prepared as above, excluding the agar setting agent, and store at RT or 4 °C after sterilization.

2.4 Hemolymph Sampling

1. Laminar flow hood.
2. Temperature-controlled centrifuge.
3. Sterile tissue (28 × 28 cm, folded twice).
4. Microscissors.
5. Ice.
6. Liquid nitrogen.
7. 1.5-mL sterile disposable microcentrifuge tubes.
8. 70% ethanol.
9. 95% ethanol.
10. 0.5% sodium hypochlorite.

2.5 Hemocyte Count

1. Inverted phase-contrast light microscope.
2. Bürker Counting Chamber.
3. 1.5-mL sterile microcentrifuge tubes.
4. Saline Solution for Lepidoptera (SSL): sucrose (210 mM), KCl (45 mM), Tris–HCl (10 mM), pH 7.0.
5. 1% formaldehyde.

2.6 Melanization Response and In Vitro Antimicrobial Activity

1. Temperature-controlled centrifuge.
2. Shaking incubator.
3. Microplate reader.
4. 1.5-mL sterile microcentrifuge tubes.
5. 96-Well sterile polystyrene plate.
6. *E. mundtii* stock solution (25% glycerol, maintained at −80 °C).
7. 25 mM phenylthiourea (PTU).
8. Ampicillin (50 mg/mL).

2.7 Morphological Analysis of the PM

1. Optical microscope and transmission electron microscope (TEM).
2. Forceps and microscissors.
3. CO_2.
4. Saline Solution for Lepidoptera (SSL) as in Subheading 2.5.
5. 0.2 M sodium cacodylate buffer (pH 7.4). Dissolve 4.28 g of sodium cacodylate trihydrate in 100 mL of water. Adjust pH with HCl before filling to the final volume.
6. 4% glutaraldehyde in 0.1 M sodium cacodylate buffer, pH 7.4. Dilute glutaraldehyde stock solution with 0.2 M sodium cacodylate buffer (*see* **Note 3**).
7. 1% osmium tetroxide in 0.1 M sodium cacodylate buffer (pH 7.4). Dilute osmium tetroxide stock solution with 0.2 M sodium cacodylate buffer.
8. 70% and 90% ethanol solutions in water; 100% ethanol.
9. Embedding resin (Epon-Araldite 812 mixture).
10. Propylene oxide/Epon-Araldite 812 mixture: mix (1:1) Epon-Araldite 812 and propylene oxide (*see* **Note 3**).
11. Lead citrate: dissolve 1.33 g of lead nitrate (MW 331.2 g/mol) and 1.87 g of sodium citrate (MW 294.10 g/mol) in 50 mL of water, and then add 8 mL NaOH 1 M.
12. Uranyl acetate. Dissolve uranyl acetate powder in water until saturation.
13. Crystal violet (1% in water) and basic fuchsin (0.13% in water; *see* **Note 4**).
14. Embedding molds.
15. 200–300 mesh copper grids (Electron Microscopy Sciences, Hatfield, USA).
16. Glass microscope slides, coverslips, Eukitt® mounting medium.

Prepare all the solutions in distilled water and store them at 4 °C until use.

3 Methods

3.1 Preparation of Germ-Free Silkworm Batches

To obtain germ-free silkworm larvae, developing eggs must be surface-sterilized and incubated in aseptic conditions before hatching. Larvae are then aseptically transferred to germ-free artificial diet and maintained for an oral infection experiment.

1. After eggs develop a white color [around 9/10 days from the beginning of incubation, at 25 ± 1 °C, in 12 h light:12 h dark (12:12 LD), 85% relative humidity (RH)], transfer them to a

laminar flow hood. Soak the eggs in 70% ethanol for 1 min and then 0.5% sodium hypochlorite for 20 min and rinse with 95% ethanol. Transfer the eggs to a sterile container and air-dry. Seal the container and incubate until hatching at 25 ± 1 °C in 12:12 L:D, 85% RH.

2. After hatching, transfer the germ-free first instar larvae to artificial diet with a sterilized paintbrush. Silkworm can remain in the same germ-free diet container until the end of the third instar (*see* **Note 5**).
3. Using a sterile forceps, the larger fourth instar larvae must be aseptically transferred to new germ-free artificial diet. After 4–5 days the fourth instar larvae will molt over 2–3 days. These silkworms must be aseptically moved to an empty sterile container and starved at 21 ± 1 °C until the required number of individuals is reached to perform subsequent experiments (*see* **Note 6**).

3.2 Oral Infection of *B. mori* with *E. mundtii*

3.2.1 Preparation of the *E. mundtii* Culture

1. Prepare an *E. mundtii* streak plate from a 25% glycerol stock and incubate overnight (ON) at 30 °C.
2. Use a single colony to inoculate 4 mL PCA broth in a 15-mL sterile tube. Incubate at 30 °C ON with vigorous shaking.
3. Measure the optical density (OD) at a wavelength of 600 nm (as an indication: an OD_{600} = 0.11 corresponds to about 2.1×10^7 *E. mundtii* cells/mL). Serially dilute the *E. mundtii* culture in PBS to a final concentration of 5×10^2 cells/mL (*see* **Notes 7** and **8**).

3.2.2 Silkworm Oral Infections with *E. mundtii*

1. Infection treatments should be performed in triplicate. Therefore three contaminated and three control containers should be prepared for each experiment. Inside a laminar flow hood, spread 2 mL of the bacterial solution evenly over the germ-free artificial diet. As a negative control, apply 2 mL of sterile PBS to sterile artificial diet containers. Close the containers.
2. Aseptically transfer germ-free fifth instar larvae to the treated artificial diets (*see* **Note 6**).
3. After the desired time of exposure, aseptically transfer the larvae to a new germ-free artificial diet container (*see* **Notes 8** and **9**).

3.3 Sampling of Hemolymph

Work inside the laminar flow hood. Before opening the silkworm containers, sterilize the external surface with 70% ethanol.

1. Collect each larva with sterilized forceps.
2. Submerge the larva in 70% ethanol for 5 s; immediately transfer to 5% sodium hypochlorite for 5 s and then to 70% ethanol for 5 s.

3. Blot dry with sterile tissue.
4. Over a sterile microcentrifuge tube, use sterile microscissors to puncture the tip of the first abdominal proleg.
5. Apply gentle pressure to drain the hemolymph into the microcentrifuge tube (*see* **Note 10**).

3.4 Measuring the Establishment of the Systemic Infection

1. Perform oral infection experiments as in Subheading 3.2.
2. Daily postinfection, aseptically extract hemolymph as in Subheading 3.3 from three/six silkworms per timepoint.
3. Serially dilute each hemolymph sample in sterile PBS (1:2, 1:10, 1:100, 1:1000).
4. Plate 5 μL from each dilution as a drop on PCA plate and tilt the plate to streak the haemolymph. Seal with parafilm.
5. Incubate the plates at 30 °C for 48 h.
6. Record the presence and number of colonies (Fig. 1b).
7. Express the number of bacterial cells per volume (e.g., μL) of undiluted hemolymph.

3.5 Hemocyte Count

1. Perform oral infection experiments as in Subheading 3.2.
2. Daily postinfection, aseptically extract hemolymph as in Subheading 3.3 (*see* **Note 11**).
3. To count hemocytes immediately after collection, dilute hemolymph 1:1 with SSL.
4. To store hemocytes long term, dilute hemolymph 1:1 with 1% formaldehyde and store at 4 °C.
5. Load 20 μL of the 1:1 hemolymph dilution (**steps 3** and **4**) into a counting chamber. Put the chamber under the microscope, and count the number of cells following the Bürker double ruling (count three to five squares). Multiply the number of hemocytes by the dilution factor, i.e., 20,000, to obtain the number of hemocytes per mL.

3.6 Rate of Melanization Response

1. Perform infection experiments as per Subheading 3.2.
2. Daily postinfection, aseptically extract hemolymph as in Subheading 3.3 using chilled microcentrifuge tubes (on ice), and maintain samples on ice (*see* **Note 12**).
3. Centrifuge the hemolymph in a prechilled centrifuge (4 °C) at 1000 RCF for 5 min, transfer the hemocyte-free plasma to a new chilled microcentrifuge tube, freeze in liquid nitrogen, and store at −20 °C until use.
4. To measure the melanization response, thaw all plasma samples on ice.

5. To a sterile polystyrene 96-well plate, add 100 μL of the plasma per well. For each plasma sample, prepare at least three wells, and perform the analysis in triplicate.
6. As internal negative controls, in parallel prepare three wells, each containing 100 μL of the same plasma sample, supplemented with 2.5 mM PTU (final concentration) to inhibit melanization.
7. Immediately measure the OD_{450} using a plate reader (time 0).
8. Repeat the readings at 450 nm at 10-min intervals for 50 min (i.e., 10, 20, 30, 40, and 50 min).
9. For each sample, calculate the ΔOD_{450} subtracting the OD_{450} at time 0 from the OD_{450} s at the other timepoints.
10. Graph the different ΔOD_{450} s versus time and include a linear regression (Fig. 2a–c).

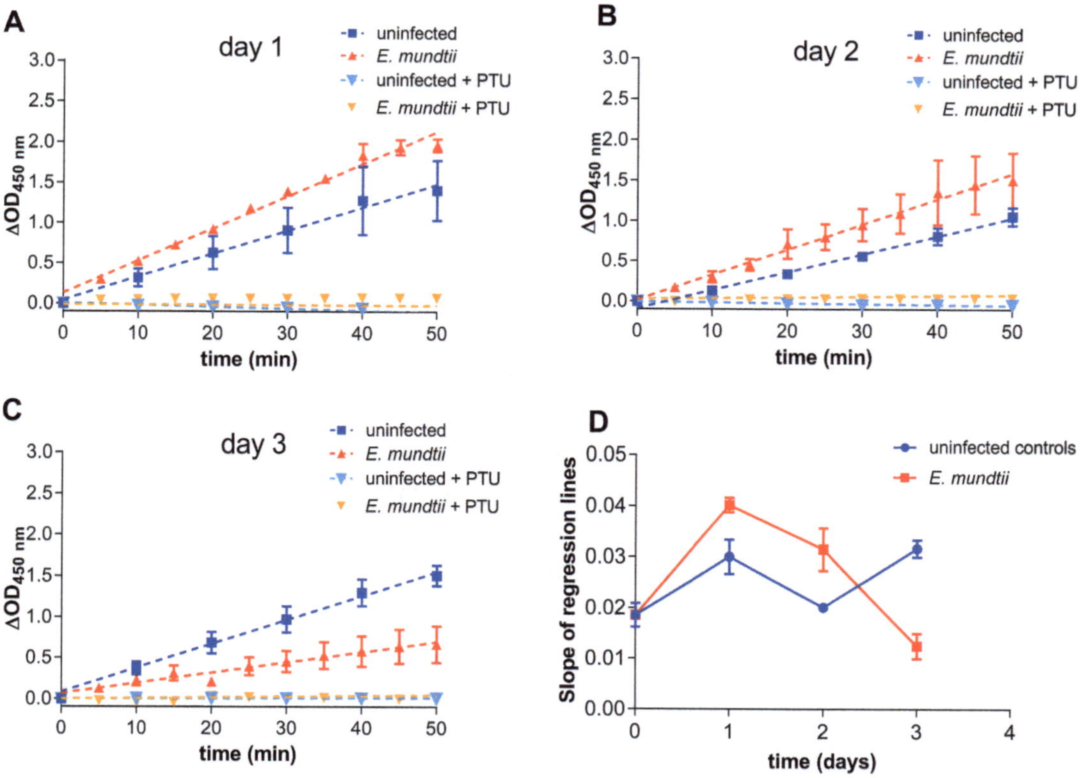

Fig. 2 Melanization rate in tropical polyvoltine silkworm plasma evaluated for 3 days in *E. mundtii*-infected larvae and in the relative uninfected controls. (**a–c**) Melanization curves and regression lines (mean ± SD, N = three replicates of three larvae each) calculated plotting ΔOD_{450} versus time for plasma samples collected at day 1 (**a**), day 2 (**b**), day 3 (**c**) postinfection. As a negative control of melanization, PTU is added at a final concentration of 2.5 mM. (**d**) Melanization rate curves obtained plotting the slopes (mean ± SD) of the regression lines obtained in (**a–c**) against the day of infection. (Data are modified from Romoli et al. (2017) [7])

11. To evaluate the melanization speed of plasma samples collected at different days after bacterial exposure, calculate the slope of each regression line, and graph these data against day postinfection (Fig. 2d).

3.7 *In Vitro* Antimicrobial Activity of Hemolymph

1. Perform infection experiments as per Subheading 3.2.
2. Daily postinfection, aseptically extract hemolymph as in Subheading 3.3 in chilled microcentrifuge tubes on ice (*see* **Note 12**).
3. To each hemolymph sample, immediately add PTU at a final concentration of 2.5 mM to prevent melanization.
4. Centrifuge the samples in a prechilled (4 °C) centrifuge at 1000 RCF for 5 min to pellet the hemocytes. Transfer the hemocyte-free plasma to a sterile microcentrifuge tube and freeze in liquid nitrogen. Store the samples at –20 °C until use.
5. Prepare a fresh culture of *E. mundtii* as per Subheading 3.2.1, and dilute to a final concentration of 5 × 10^5 CFU/mL in PCA broth.
6. Thaw all plasma samples on ice.
7. To a sterile polystyrene 96-well plate, add 25 μL of the hemocyte-free plasma per well. For each plasma sample, perform the analysis in triplicate, preparing at least three wells.
8. To each plasma-containing well, add 200 μL of the *E. mundtii* culture.
9. Always in triplicate, prepare a negative growth control adding 25 μL of ampicillin to a final concentration of 50 μg/mL to 200 μL of the *E. mundtii* culture.
10. Always in triplicate, prepare a positive growth control adding 25 μL of sterile PBS to 200 μL of the *E. mundtii* culture.
11. Immediately measure the OD_{600} on a plate reader.
12. Incubate the plate at 30 °C with soft shaking (50 RPM), and measure the OD_{600} at 3-, 5-, 7-, 9-, and 24-h timepoints using the plate reader.
13. Log transform the OD_{600} values.
14. Calculate the bacterial growth rate (μ), using the formula:

$$\mu = 2.303 \times \left(\lg \mathrm{OD}_{t_n} - \lg \mathrm{OD}_{t_0}\right) / \left(t_n - t_0\right)$$

with t_n indicating the different timepoints and t_0 the initial time 0, as described in [7, 14].

15. Plot μ data against time, obtaining the *E. mundtii* growth rate in each plasma sample and in both negative and positive controls (Fig. 3a).
16. To compare the growth of *E. mundtii* treated with different plasma samples, for each μ curve, calculate the area under the curve in the time span comprised between 3 and 9 h

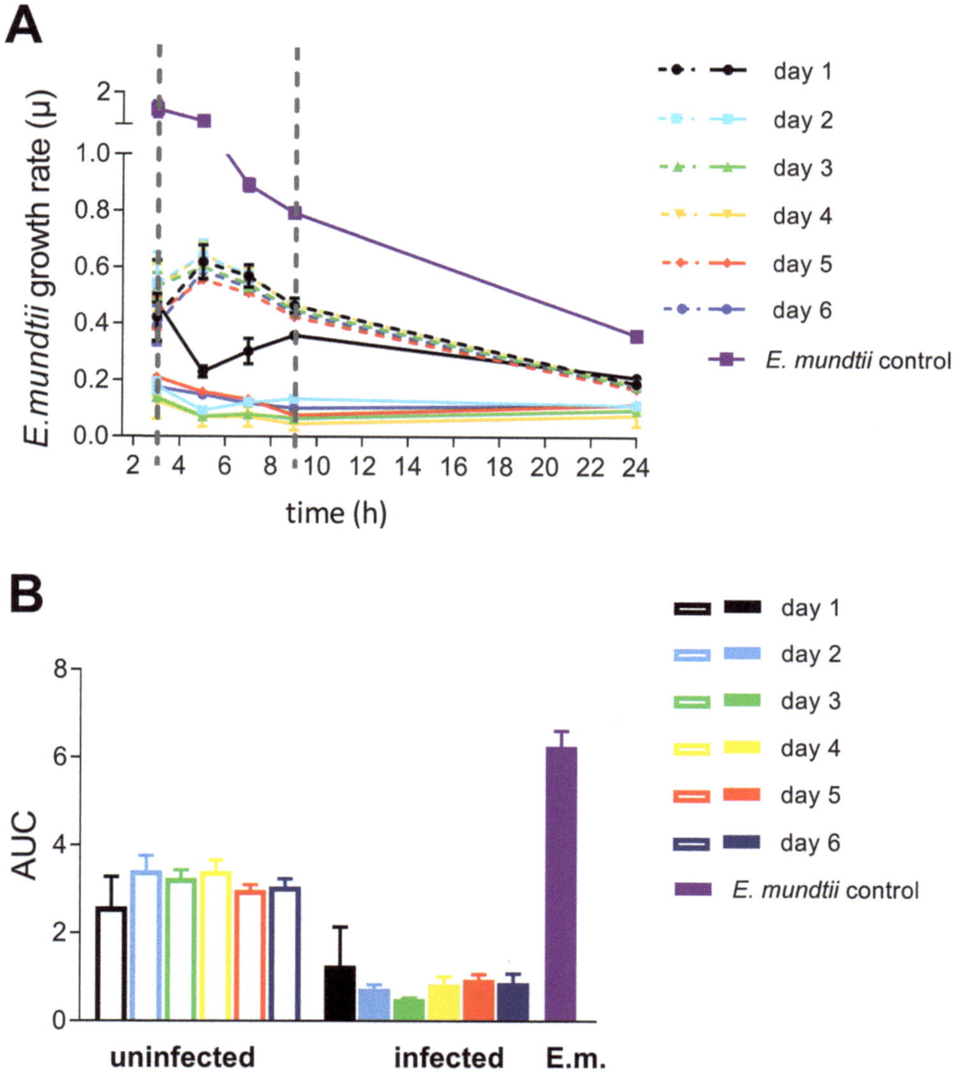

Fig. 3 Antimicrobial activity kinetics in tropical polyvoltine silkworm plasma samples evaluated for 6 days in *E. mundtii*-infected larvae and in the relative uninfected controls. (**a**) In vitro growth rate μ, mean ± SEM, *N* = three replicates of three larvae each) of *E. mundtii* (5 × 10^5 CFU/mL) in the presence of hemocyte-free plasma sampled from infected larvae (solid lines) and uninfected controls (dashed lines) for 6 days after *E. mundtii* oral exposure. Purple line: *E. mundtii* control. Vertical gray dashed lines indicate the timepoints between 3 and 9 h *post-inoculum* used to calculate the area under the curve (AUC). (**b**) AUC (mean ± SEM): Area under μ curves calculated for each condition in the timepoints between 3 and 9 h *post-inoculum*. (Data are modified from Romoli et al. (2017) [7])

post-inoculum, corresponding to the period of time in which *E. mundtii* shows the highest growth rate values in culture media (Fig. 3b).

3.8 Morphological Analysis of the PM

1. Anesthetize larvae with CO_2 for a few seconds.
2. Dissect the larva by cutting the dorsal integument and collect the midgut (*see* **Note 13**).
3. Wash the midgut extensively with SSL to remove hemocytes and fat body debris.
4. Put the midgut into a 1.5-mL plastic tube filled with 4% glutaraldehyde in 0.1 M sodium cacodylate buffer, and fix the sample ON at 4 °C (*see* **Note 14**).
5. Wash thoroughly with 0.1 M cacodylate buffer (three to five washes, 5–10 min each) to remove residues of fixative.
6. Postfix the sample with 1% osmium tetroxide for 1 h.
7. Wash thoroughly with 0.1 M cacodylate buffer (three to five washes, 5–10 min each) to remove residual osmium tetroxide.
8. Dehydrate in an ethanol series by incubating the samples in 70%, 90%, and 100% ethanol for 30 min each (*see* **Note 15**).
9. Incubate the specimen in propylene oxide/Epon-Araldite 812 mixture for 1 h (*see* **Note 16**).
10. After removing the propylene oxide/Epon-Araldite 812 mixture, add Epon-Araldite 812 mixture to the samples and incubate ON at RT.
11. Transfer the midgut samples to embedding molds, cover them with fresh resin, and allow to polymerize ON at 70 °C.
12. Cut sections (0.7-μm-thick) with a microtome and collect them on glass slides. Stain them with basic fuchsin and crystal violet for a second each. Mount the coverslip with Eukitt.
13. Cut ultrathin sections (70-nm-thick) with an ultramicrotome and collect them on grids. Staining procedure: 4-min lead citrate, washes with water, 8-min uranyl acetate, washes with water, 4-min lead citrate, and washes with water.
14. Analyze specimens using an optical microscope (for slides) or TEM (for grids) (Fig. 4).

4 Notes

1. Various artificial diet formulations, hydration rates, and cooking/sterilization conditions have been rigorously and systematically tested, and the reported conditions provide optimal development of silkworms: patented method (WIPO code

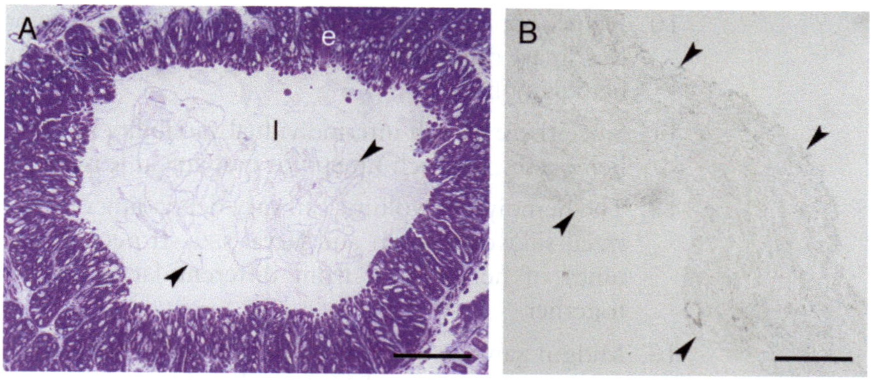

Fig. 4 Images of silkworm peritrophic matrix. (**a**) Optical microscopy, (**b**) transmission electron microscopy. Arrowheads indicate the peritrophic matrix; e, midgut epithelium; l, lumen. Bar, 200 μm (**a**), 500 nm (**b**)

10MI2013A000596). Diet powder can be stored short term at RT and shielded from direct light, or for 5–6 months at −20 °C.

2. Wrap wads of lab tissue in aluminum foil and autoclave, and dry at 160 °C for 2 h or at 60 °C overnight.
3. This solution must be freshly prepared.
4. The solution must be filtered with filter paper before use.
5. Given their small size and the small quantity of food they require, hundreds of first to third instar larvae can be maintained in the same container.
6. To maintain optimum growth conditions, we find that a maximum of 15 fifth instar larvae should be housed per diet container with a size of 30 × 16 × 8 cm.
7. It is a good practice to check the bacterial concentrations by plating 100 μL of the bacterial culture (5×10^2 cells/mL) used to infect silkworms on a PCA plate and incubating ON. About 50 CFU/plate are expected.
8. Silkworm strains show variable sensitivity to *E. mundtii* infection. If the sensitivity to *E. mundtii* of the selected silkworm strain is unknown, pilot infection experiments should be performed testing at least three different bacterial concentrations across orders of magnitude. In addition, the optimal exposure time may need to be optimized. As an indication, we find from 6 to 24 h exposure times effective for oral infections with *E. mundtii* at a concentration of 5×10^2 cells/mL.
9. To determine the quantity of the ingested food, weigh each container before and after the bacterial exposure. Remove frass before reweighting each container after bacterial exposure.

10. Do not apply too much pressure when draining hemolymph as this may cause the gut to rupture and contaminate the hemolymph.
11. Since there is high interindividual variability in the number of hemocytes, for each timepoint evaluate at least ten larvae.
12. The hemolymph volume can vary depending on both the analyzed silkworm strain and larval size. If necessary, equal volumes of hemolymph from different larvae can be pooled together.
13. Midgut samples must be collected using particular attention to preserve the inner PM. In order to better allow penetration of fixative into the tissue, midgut samples should not be longer than 2 mm.
14. The samples can be stored in the fixative at 4 °C until further processing.
15. If necessary, the samples can be stored in 70% ethanol at 4 °C until further processing.
16. Propylene oxide/Epon-Araldite 812 mixture must be completely removed from the plastic tube to avoid softening of the resin during the subsequent embedding step.

Acknowledgments

The authors acknowledge CARIPARO (Progetti di Eccellenza 2011/2012) for support to S.C., F.S., and G.T.; F.S. also thanks a grant from Università degli Studi di Padova (CPDA154301).

References

1. Guo-Ping K, Xi-Jie G (2011) Overview of silkworm pathology in China. Afr J Biotechnol 10(79):18046–18056
2. Brey PT, Lee W-J, Yamakawa M, Koizumi Y, Perrot S, Francois M et al (1993) Role of the integument in insect immunity: epicuticular abrasion and induction of cecropin synthesis in cuticular epithelial cells. Proc Natl Acad Sci U S A 90(13):6275–6279
3. Hegedus D, Erlandson M, Gillott C, Toprak U (2009) New insights into peritrophic matrix synthesis, architecture, and function. Annu Rev Entomol 54:285–302
4. Chen K, Lu Z (2018) Immune responses to bacterial and fungal infections in the silkworm, *Bombyx mori*. Dev Comp Immunol 83:3–11
5. Wu K, Yang B, Huang W, Dobens L, Song H, Ling E (2016) Gut immunity in Lepidopteran insects. Dev Comp Immunol 64:65–74
6. Tanaka H, Yamakawa M (2011) Regulation of the immune responses in the silkworm, *Bombyx mori*. Invertebr Surviv J 8:59–69
7. Romoli O, Saviane A, Bozzato A, D'Antona P, Tettamanti G, Squartini A et al (2017) Differential sensitivity to infections and antimicrobial peptide-mediated immune response in four silkworm strains with different geographical origin. Sci Rep 7(1):1048
8. Nagaraju J, Klimenko V, Couble P, Reeves E (2000) The silkworm *Bombyx mori*, a model genetic system. In: Encyclopedia of genetics. Fitzroy Dearborn Publishers, Chicago, p 219239
9. Sun Z, Lu Y, Zhang H, Kumar D, Liu B, Gong Y et al (2016) Effects of BmCPV infection on silkworm *Bombyx mori* intestinal bacteria. PLoS One 11(1):e0146313

10. Nishida S, Ishii M, Nishiyama Y, Abe S, Ono Y, Sekimizu K (2017) *Lactobacillus paraplantarum* 11-1 isolated from rice bran pickles activated innate immunity and improved survival in a silkworm bacterial infection model. Front Microbiol 8:436
11. Alebiosu IB, Olatunde GO, Pitan OOR (2013) Variations in some traits of two silkworm *Bombyx mori* L. (Lepidoptera: Bombycidae) hybrids fed on mulberry leaves of different maturity stages. Nig J Sci 47:1–7
12. Gangwar SK (2010) Impact of varietal feeding of eight mulberry varieties on *Bombyx mori* L. Agric Biol J N Am 1(3):350–354
13. Cappellozza S, Saviane A, Tettamanti G, Squadrin M, Vendramin E, Paolucci P et al (2011) Identification of *Enterococcus mundtii* as a pathogenic agent involved in the "flacherie" disease in *Bombyx mori* L. larvae reared on artificial diet. J Invertebr Pathol 106(3):386–393
14. Widdel F (2007) Theory and measurement of bacterial growth. Grundpraktikum Mikrobiologie 4(11):1–11

Chapter 15

Silkworm Infection Model for Evaluating Pathogen Virulence

Yasuhiko Matsumoto and Kazuhisa Sekimizu

Abstract

Animal infection experiments are necessary to elucidate host-pathogen interactions in infectious diseases. Both basic and applied studies of infectious diseases can be performed not only in mammals but also invertebrates. Here we describe a silkworm infection model that is useful for evaluating pathogen virulence. The silkworm, an invertebrate, has several advantages for large-scale in vivo screening to identify pathogen virulence factors as well as host factors that inhibit pathogen virulence. We present a basic technique for studying host-pathogen interactions in a silkworm *Staphylococcus aureus* infection model.

Key words Silkworm, Infection, Host-pathogen interaction, Injection, *Staphylococcus aureus*

1 Introduction

Pathogenic microorganisms infect humans and cause various infectious diseases. Elucidating the pathogenic mechanisms of infectious diseases caused by pathogenic microorganisms is necessary for establishing therapeutic and preventive methods. Basic research using animal models mimicking human infectious diseases is indispensable for this purpose.

Several pathogenic infection models using various mammals have been proposed [1–3]. The use of mammalian models is costly and complicated. To overcome these issues, infection models using invertebrates such as the fruit fly *Drosophila melanogaster*, the nematode *Caenorhabditis elegans*, and the greater wax moth *Galleria mellonella* have been developed [4, 5]. Invertebrate animal models have advantages compared with mammalian models, such as (1) lower breeding costs, (2) larger numbers of individuals can be reared in a smaller space, (3) fewer ethical problems, and (4) tenfold smaller amounts of samples compared to mice are needed due to smaller body size [6].

Federica Sandrelli and Gianluca Tettamanti (eds.), *Immunity in Insects*, Springer Protocols Handbooks,
https://doi.org/10.1007/978-1-0716-0259-1_15, © Springer Science+Business Media, LLC, part of Springer Nature 2020

The silkworm, *Bombyx mori*, is proposed as an experimental model animal for pathogenic microorganisms infecting humans [7–12]. Silkworm rearing procedures are well established due to the long history of sericulture. In contrast to mammals, body temperature of the silkworm can be easily controlled by changing the rearing temperature [6]. By simple intra-midgut injection, silkworms can be used to investigate the pathogenicity of bacteria infecting the intestinal tract. These features of the silkworm as an experimental animal facilitate studies of the molecular mechanisms underlying host-pathogen interactions.

Silkworm infection models are established for human pathogens listed below: Gram-negative bacteria such as *Pseudomonas aeruginosa* [7], pathogenic *Escherichia coli* [8], *Serratia marcescens* [9], and *Vibrio cholerae* [7]; Gram-positive bacteria such as *Staphylococcus aureus* [7], *Streptococcus pyogenes* [10], and *Listeria monocytogenes* [11]; and fungi such as *Candida albicans* [12], *Candida tropicalis* [12], *Candida glabrata* [13], *Cryptococcus neoformans* [14], *Aspergillus fumigatus* [15], *Arthroderma vanbreuseghemii* [16], *Arthroderma benhamiae* [16], *Microsporum canis* [16], *Trichophyton rubrum* [16], and *Rhizopus oryzae* [17]. Silkworms are also killed by injection of extracellular toxins, such as α-toxin and β-toxin of *S. aureus*, exotoxin A of *Pseudomonas aeruginosa*, diphtheria toxin, and hemolysin of *Bacillus cereus* [18, 19]. The silkworm infection model can be used to quantitatively measure pathogen virulence based on the LD_{50} that is determined as the number of pathogen injected cells required to kill 50% of the silkworms.

The silkworm infection models are applicable for the identification of pathogenic genes of pathogens [6, 20, 21]. In case of *S. aureus*, avirulent mutants were screened by using a silkworm infection model from a gene-deficient mutant library of *S. aureus* [10]. The mutants also exhibited lower pathogenicity in a mouse infection model as well [10]. These genes, *cvfA*, *cvfB*, and *cvfC*, regulate production of *S. aureus* virulence factors, such as hemolysins, proteases, and nucleases [10]. Therefore, the strategy using silkworm infection models with gene-disrupted mutants of a human pathogen is useful for the identification of pathogenic genes.

The silkworm infection models are also applicable for the identification of candidates of anti-bacterial, anti-fungal, and anti-viral drugs [22, 23]. In case of *S. aureus*, the death of silkworms infected with *S. aureus* is cured by administration of anti-bacterial drugs, such as chloramphenicol [7]. Candidates were screened by using a silkworm infection model with *S. aureus* from a natural product library. Lysocin E included in the candidates was identified as a novel antibiotic for treatment of *S. aureus* infection [24]. Therefore, the strategy using silkworm infection models with a chemical or natural product library is useful for the identification of antibiotics for human pathogens.

Here, we outline a method for evaluating the pathogenicity of *S. aureus* using the silkworm infection model.

2 Materials

2.1 Reagents, Bacteria, Insects, and Equipment

1. Tryptic soy broth (TSB) (Becton, Dickinson and Company, MD, USA).
2. Saline (0.9% NaCl) (Otsuka, Tokushima, Japan).
3. *S. aureus* Newman strain (Public Health England, Salisbury, UK).
4. Silkworm larvae (Hu·Yo × Tukuba·Ne) (Ehime Sansyu, Ehime, Japan).
5. Artificial diet for rearing silkworms (Silkmate without antibiotic: Sysmex, Hyogo, Japan).
6. Paper (Kimwiper: Kracie, Tokyo, Japan).
7. Plastic packs (Denka polymer, Tokyo, Japan).
8. 1-mL syringe with 27-gauge needle (Terumo syringe: SS-01T2719S, Terumo, Japan).
9. 2-mL tubes (Safe-Lock Tubes 2.0 mL: Eppendorf, Hamburg, Germany).
10. 50-mL tubes (Falcon polypropylene conical tube, Corning Life Sciences, Tamaulipas, Mexico).
11. Shaker (BR-40LF: TAITEC, Aichi, Japan).
12. Centrifuge (CAX-371, Tomy, Tokyo, Japan).
13. Incubator (MIR-154S-PJ: Panasonic, Osaka, Japan).

2.2 Preparation of Silkworm

1. Purchase fifth instar silkworm larvae (Hu·Yo × Tukuba·Ne) from a commercial supplier (Ehime Sansyu, Ehime, Japan). Fifth instar larvae are used for infection experiments (*see* **Note 1**).
2. Feed diet (Silkmate without antibiotic: Sysmex, Hyogo, Japan) to fifth instar larvae at 27 °C for 1 day (*see* **Note 2**).
3. After 1 day feeding, the silkworms are used in the infection experiment (*see* Fig. 1a).

2.3 Established Silkworm Infection Models

Silkworm infection models are established for human pathogens listed below:

1. Gram-negative bacteria

   Pseudomonas aeruginosa, pathogenic *Escherichia coli*, *Serratia marcescens*, and *Vibrio cholerae*.
2. Gram-positive bacteria

   Staphylococcus aureus, *Streptococcus pyogenes*, and *Listeria monocytogenes*.

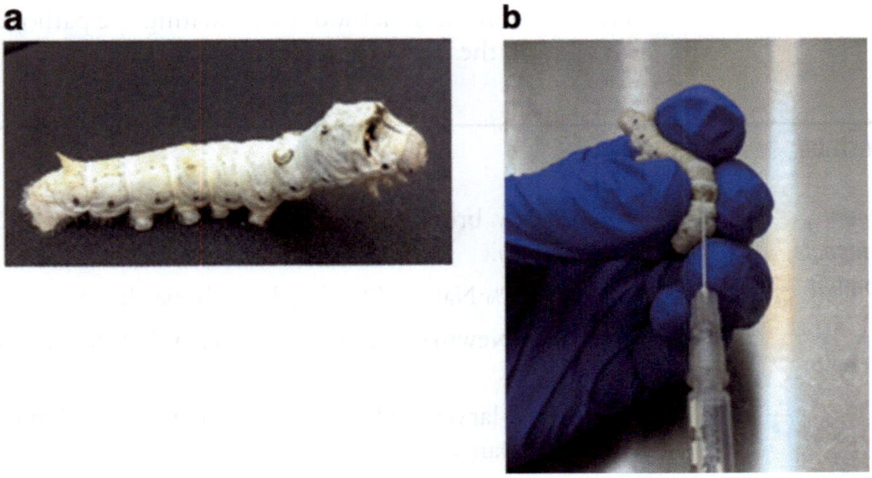

Fig. 1 Injection of bacterial solution into the silkworm hemolymph. (**a**) A fifth instar silkworm fed an artificial diet for 1 day. (**b**) Bacterial solution is injected into the silkworm hemolymph

3. Fungi

   Candida albicans, *Candida tropicalis*, *Candida glabrata*, *Cryptococcus neoformans*, *Aspergillus fumigatus*, *Arthroderma vanbreuseghemii*, *Arthroderma benhamiae*, *Microsporum canis*, *Trichophyton rubrum*, and *Rhizopus oryzae*.

3 Methods

3.1 Preparation of Bacterial Solutions of S. aureus

1. Streak *S. aureus* Newman strain (Public Health England, Salisbury, UK) from −80 °C glycerol stock onto agar plates of tryptic soy broth (TSB) (Becton, Dickinson and Company, MD, USA) (*see* **Note 3**).
2. Incubate the plate at 37 °C for 24 h.
3. Inoculate a single colony from a TSB agar plate into 5 mL of TSB in a 50-mL tube (Falcon polypropylene conical tube, Corning Life Sciences, Tamaulipas, Mexico).
4. Incubate the tube at 37 °C for 24 h with shaking at 160–200 rpm (BR-40LF: TAITEC, Aichi, Japan).
5. Dilute full-growth bacterial culture 1000-fold with TSB. Five microliters of full-growth culture are added into 5 mL of TSB in a 50-mL tube.
6. Incubate at 37 °C in a shaking incubator at 160–200 rpm for 18–24 h.
7. Transfer 1.5 mL of bacterial culture (OD_{600} 5–10) to a 2-mL tube (Safe-Lock Tubes 2.0 mL: Eppendorf, Hamburg, Germany) (*see* **Note 4**).

8. Centrifuge the 2-mL tube for 5 min at 8,000 × *g* (CAX-371, Tomy, Tokyo, Japan), and remove the supernatant by aspiration.
9. Add 1.5 mL of saline (Otsuka, Tokushima, Japan) to the 2-mL tube with the bacterial pellet and suspend by pipetting.
10. Serially dilute twofold with saline (Otsuka, Tokushima, Japan). Serially twofold diluted bacterial solutions (750 μL) are prepared as follows. An aliquot (750 μL) of the bacterial solution is added to 750 μL of saline in a 2-mL tube and mixed. An aliquot (750 μL) of the twofold diluted bacterial solution is added to 750 μL of saline in a 2-mL tube.
11. Determine viable cell number (colony-forming units [CFU]/mL) of the bacterial solution using the colony counting method [25]. The bacterial solution is diluted with saline 10^6-fold. One hundred microliters of the diluted solution are spread on a TSB agar plate. The plate is incubated at 37 °C for 1 day. Bacterial colony number on the plate is counted.

3.2 Injection of Silkworms

1. Fill *S. aureus* bacterial solution or control solution (saline) into 1-mL syringe with 27-gauge needle (Terumo syringe: SS-01T2719S, Terumo, Japan) (*see* **Note 5**).
2. Administer 50 μL of *S. aureus* bacterial solution into the hemolymph by injecting the silkworm dorsally (*see* **Note 6**) (*see* Fig. 1b) [6].

3.3 Maintenance of Infected Silkworms

1. Place silkworms injected with *S. aureus* on paper (Kimwiper: Kracie, Tokyo, Japan) in plastic packs (Denka polymer, Tokyo, Japan) (*see* Fig. 2a).
2. Cover the plastic pack.
3. Place the plastic pack containing the infected silkworms into an incubator (MIR-154S-PJ: Panasonic, Osaka, Japan) at 27 °C (*see* Fig. 2b). Do not feed diet to silkworms after injecting the bacterial solution.

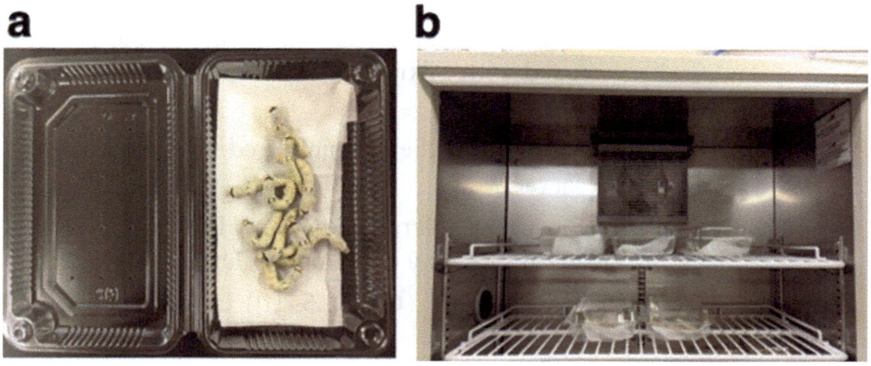

Fig. 2 Incubation of silkworms infected with *S. aureus*. (**a**) Silkworms infected with *S. aureus* in a plastic pack. (**b**) The packs in an incubator

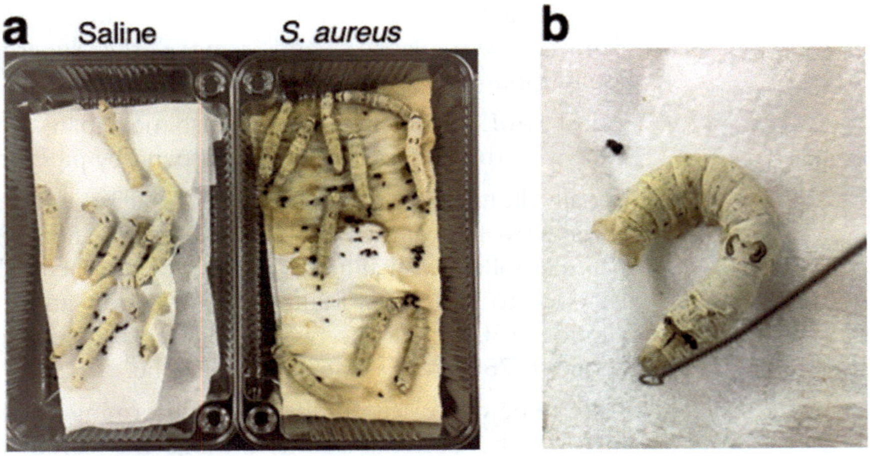

Fig. 3 Death of silkworms infected by *S. aureus*. (**a**) Picture shows that silkworms died after injection of *S. aureus*. Fifty microliters of saline or bacterial solution (3 × 10^9 CFU/mL) of *S. aureus* Newman strain was injected into the silkworm hemolymph. Photograph was taken after incubation at 27 °C for 24 h. Left, saline injected; right, *S. aureus* (Newman strain) injected. (**b**) Check survival of silkworms using an inoculating loop. Live silkworms move when the silkworm head is touched with an inoculating loop

3.4 Determination of Silkworm Survival

1. Open the plastic pack with the infected silkworms in a safety cabinet.
2. Touch the head of the infected silkworms with an inoculating loop to check the reaction (*see* **Note 7**) (*see* Fig. 3).
3. Count the number of surviving silkworms at least twice per day (*see* **Note 8**) (*see* Fig. 4). The number of surviving silkworms was counted at 24 h after injection for determination of LD_{50}.

4 Notes

1. The rearing method of fifth instar larvae from eggs was described previously [7]. Fertilized silkworm eggs were obtained from Ehime Sansyu (Ehime, Japan). Hatched larvae were fed with Silkmate 2S (Ehime Sansyu, Ehime, Japan) at 27 °C.
2. The amount of diet administered to the silkworms was 1.5 g per larvae.
3. Perform in a safety cabinet (MHE-130AJ: SANYO, Osaka, Japan) at biosafety containment level 2 (BSL-2).
4. Bacterial culture (OD_{600} 5–10) corresponds to 5–10 × 10^9 CFU/mL.

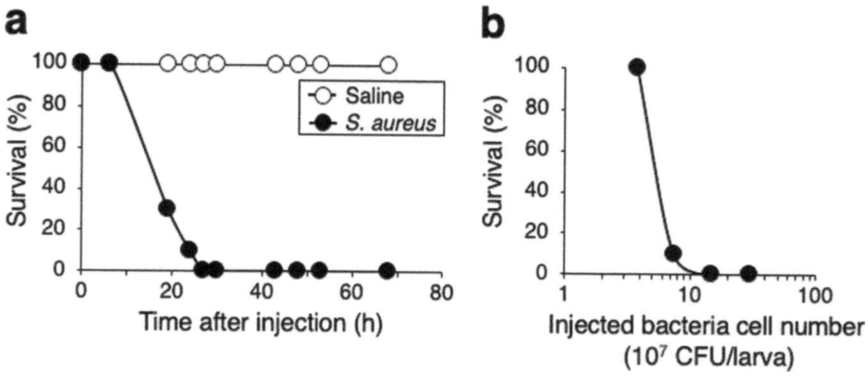

Fig. 4 Evaluation of virulence of *S. aureus* in the silkworm infection model. (**a**) Time course of survival of silkworms after injection of *S. aureus*. Fifty microliters of saline or bacterial solution (3×10^9 CFU/mL) of *S. aureus* Newman strain was injected into the silkworm hemolymph. The silkworms were incubated at 27 °C. Surviving number of silkworms was counted for 3 days. $N = 10$ per group. (**b**) Dose-response of *S. aureus* for killing silkworms. Fifty microliters of saline or bacterial solutions (0.75, 1.5, 3, and 6×10^9 CFU/mL) of *S. aureus* Newman strain was injected into the silkworm hemolymph. The silkworms were incubated at 27 °C for 24 h and the number of surviving silkworms was counted. The *x*-axis shows injected bacterial number per silkworm larva. $N = 10$ per group

5. The injection experiment should be conducted in a safety cabinet (MHE-130AJ: SANYO, Osaka, Japan) at biosafety containment level 2 (BSL-2).
6. Fill the syringe with 50 μL of sample solution and inject the entire volume into hemolymph of silkworms.
7. When head and/or legs of the silkworm move, it is judged as alive. When head and/or legs do not move, it is judged as dead.
8. In Fig. 4a, the survival number of the silkworms was checked at 0, 6, 19, 24, 27, 30, 43, 48, 53, and 68 h after injection of samples. In Fig. 4b, the survival number of the silkworms was checked at 24 h after injection of samples. If the silkworm does not die, the amounts of bacterial cells administered must be increased and the silkworm incubation conditions adjusted.

Acknowledgments

This work was supported by JSPS KAKENHI grant number JP15H05783 (Scientific Research (S) to K.S.) and JSPS KAKENHI grant number JP17K08288 (Scientific Research (C) to Y.M.).

References

1. Cambier L, Heinen MP, Mignon B (2017) Relevant animal models in dermatophyte research. Mycopathologia 182:229–240
2. Lewis RE, Verweij PE (2017) Animal models for studying triazole resistance in *Aspergillus fumigatus*. J Infect Dis 216:S466–S473
3. Segal E, Frenkel M (2018) Experimental *in vivo* models of candidiasis. J Fungi 4:21
4. Elkabti A, Issi L, Rao R (2018) *Caenorhabditis elegans* as a model host to monitor the Candida infection processes. J Fungi 4:123
5. Ishii M, Matsumoto Y, Sekimizu K (2016) Usefulness of silkworm as a host animal for understanding pathogenicity of *Cryptococcus neoformans*. Drug Discov Therap 10:9–13
6. Matsumoto Y, Sekimizu K (2019) Silkworm as an experimental animal to research for fungal infections. Microbiol Immunol 63:41–50
7. Kaito C, Akimitsu N, Watanabe H et al (2002) Silkworm larvae as an animal model of bacterial infection pathogenic to humans. Microb Pathog 32:183–190
8. Miyashita A, Iyoda S, Ishii K et al (2012) Lipopolysaccharide O-antigen of enterohemorrhagic *Escherichia coli* O157:H7 is required for killing both insects and mammals. FEMS Microbiol Lett 333:59–68
9. Ishii K, Adachi T, Imamura K et al (2012) *Serratia marcescens* induces apoptotic cell death in host immune cells via a lipopolysaccharide- and flagella-dependent mechanism. J Biol Chem 287:36582–36592
10. Kaito C, Kurokawa K, Matsumoto Y et al (2005) Silkworm pathogenic bacteria infection model for identification of novel virulence genes. Mol Microbiol 56:934–944
11. Castillo Y, Suzuki J, Watanabe K et al (2016) Effect of vitamin A on *Listeria monocytogenes* infection in a silkworm model. PLoS One 11:e0163747
12. Hamamoto H, Kurokawa K, Kaito C et al (2004) Quantitative evaluation of the therapeutic effects of antibiotics using silkworms infected with human pathogenic microorganisms. Antimicrob Agents Chemother 48:774–779
13. Ueno K, Matsumoto Y, Uno J et al (2011) Intestinal resident yeast *Candida glabrata* requires Cyb2p-mediated lactate assimilation to adapt in mouse intestine. PLoS One 6:e24759
14. Matsumoto Y, Miyazaki S, Fukunaga DH et al (2012) Quantitative evaluation of cryptococcal pathogenesis and antifungal drugs using a silkworm infection model with *Cryptococcus neoformans*. J Appl Microbiol 112:138–146
15. Nakamura I, Kanasaki R, Yoshikawa K et al (2017) Discovery of a new antifungal agent ASP2397 using a silkworm model of *Aspergillus fumigatus* infection. J Antibiot 70:41–44
16. Ishii M, Matsumoto Y, Yamada T et al (2017) An invertebrate infection model for evaluating anti-fungal agents against dermatophytosis. Sci Rep 7:12289
17. Tominaga T, Uchida R, Koyama N et al (2018) Anti-rhizopus activity of tanzawaic acids produced by the hot spring-derived fungus *Penicillium* sp. BF-0005. J Antibiot 71:626–632
18. Hossain MS, Hamamoto H, Matsumoto Y et al (2006) Use of silkworm larvae to study pathogenic bacterial toxins. J Biochem 140:439–444
19. Usui K, Miyazaki S, Kaito C et al (2009) Purification of a soil bacteria exotoxin using silkworm toxicity to measure specific activity. Microb Pathog 46:59–62
20. Kaito C, Sekimizu K (2007) A silkworm model of pathogenic bacterial infection. Drug Discov Therap 1:89–93
21. Kaito C (2016) Understanding of bacterial virulence using the silkworm infection model. Drug Discov Therap 10:30–33
22. Ishii M, Matsumoto Y, Nakamura I et al (2017) Silkworm fungal infection model for identification of virulence genes in pathogenic fungus and screening of novel antifungal drugs. Drug Discov Therap 11:1–5
23. Panthee S, Paudel A, Hamamoto H et al (2017) Advantages of the silkworm as an animal model for developing novel antimicrobial agents. Front Microbiol 8:373
24. Hamamoto H, Urai M, Ishii K et al (2015) Lysocin E is a new antibiotic that targets menaquinone in the bacterial membrane. Nat Chem Biol 11:127–133
25. Ishii M, Matsumoto Y, Sekimizu K (2018) Estimation of lactic acid bacterial cell number by DNA quantification. Drug Discov Therap 12:88–91

INDEX

Federica Sandrelli and Gianluca Tettamanti (eds.), *Immunity in Insects*, Springer Protocols Handbooks,
https://doi.org/10.1007/978-1-0716-0259-1, © Springer Science+Business Media, LLC, part of Springer Nature 2020

M

N

O

P

MIX
Papier aus verantwortungsvollen Quellen
Paper from responsible sources
FSC® C105338

If you have any concerns about our products,
you can contact us on
ProductSafety@springernature.com
In case Publisher is established outside the EU,
the EU authorized representative is:
Springer Nature Customer Service Center GmbH
Europaplatz 3, 69115 Heidelberg, Germany
Printed by Libri Plureos GmbH
in Hamburg, Germany